T0350914

Parallel Computing for Data Science

With Examples in R, C++ and CUDA

Chapman & Hall/CRC
The R Series

Series Editors

Aims and Scope

This book series reflects the recent rapid growth in the development and application of R, the programming language and software environment for statistical computing and graphics. R is now widely used in academic research, education, and industry. It is constantly growing, with new versions of the core software released regularly and more than 6,000 packages available. It is difficult for the documentation to keep pace with the expansion of the software, and this vital book series provides a forum for the publication of books covering many aspects of the development and application of R.

The scope of the series is wide, covering three main threads:

- Applications of R to specific disciplines such as biology, epidemiology, genetics, engineering, finance, and the social sciences.
- Using R for the study of topics of statistical methodology, such as linear and mixed modeling, time series, Bayesian methods, and missing data.
- The development of R, including programming, building packages, and graphics.

The books will appeal to programmers and developers of R software, as well as applied statisticians and data analysts in many fields. The books will feature detailed worked examples and R code fully integrated into the text, ensuring their usefulness to researchers, practitioners and students.

Published Titles

Stated Preference Methods Using R, *Hideo Aizaki, Tomoaki Nakatani, and Kazuo Sato*

Using R for Numerical Analysis in Science and Engineering, *Victor A. Bloomfield*

Event History Analysis with R, *Göran Broström*

Computational Actuarial Science with R, *Arthur Charpentier*

Statistical Computing in C++ and R, *Randall L. Eubank and Ana Kupresanin*

Reproducible Research with R and RStudio, *Christopher Gandrud*

Introduction to Scientific Programming and Simulation Using R, Second Edition, *Owen Jones, Robert Maillardet, and Andrew Robinson*

Nonparametric Statistical Methods Using R, *John Kloke and Joseph McKean*

Displaying Time Series, Spatial, and Space-Time Data with R, *Oscar Perpiñán Lamigueiro*

Programming Graphical User Interfaces with R, *Michael F. Lawrence and John Verzani*

Analyzing Sensory Data with R, *Sébastien Lê and Theirry Worch*

Parallel Computing for Data Science: With Examples in R, C++ and CUDA, *Norman Matloff*

Analyzing Baseball Data with R, *Max Marchi and Jim Albert*

Growth Curve Analysis and Visualization Using R, *Daniel Mirman*

R Graphics, Second Edition, *Paul Murrell*

Data Science in R: A Case Studies Approach to Computational Reasoning and Problem Solving, *Deborah Nolan and Duncan Temple Lang*

Multiple Factor Analysis by Example Using R, *Jérôme Pagès*

Customer and Business Analytics: Applied Data Mining for Business Decision Making Using R, *Daniel S. Putler and Robert E. Krider*

Implementing Reproducible Research, *Victoria Stodden, Friedrich Leisch, and Roger D. Peng*

Graphical Data Analysis with R, *Antony Unwin*

Using R for Introductory Statistics, Second Edition, *John Verzani*

Advanced R, *Hadley Wickham*

Dynamic Documents with R and knitr, *Yihui Xie*

Parallel Computing for Data Science

With Examples in R, C++ and CUDA

Norman Matloff
University of California, Davis
USA

CRC Press is an imprint of the
Taylor & Francis Group, an **informa** business
A CHAPMAN & HALL BOOK

First published 2015 by Chapman & Hall

Published 2019 by CRC Press
Taylor & Francis Group
6000 Broken Sound Parkway NW, Suite 300
Boca Raton, FL 33487-2742

ISBN 13: 978-1-4665-8701-4 (hbk)

Visit the Taylor & Francis Web site at
http://www.taylorandfrancis.com

and the CRC Press Web site at
http://www.crcpress.com

Contents

Preface

Thank you for your interest in this book. I've very much enjoyed writing it, and I hope it turns out to become very useful to you. To set the stage, there are a few general points of information I wish to present.

Goals:

This book hopefully will live up to its title—*Parallel Computing for Data Science.* Unlike almost every other book I'm aware of on parallel computing, you will not find a single example here dealing with solving partial differential equations and other applications to physics. This book really is devoted to applications in data science—whether you define that term to be statistics, data mining, machine learning, pattern recognition, analytics, or whatever.[1]

This means more than simply that the book's examples involve applications chosen from the data science field. It also means that the data structures, algorithms and so on reflect this orientation. This will range from the classic "n observations, p variables" matrix format to time series to network graph models to various other structures common in data science.

While the book is chock full of examples, it aims to emphasize general principles. Accordingly, after presenting an introductory code example in Chapter 1 (general principles are meaningless without real examples to tie them to), I devote Chapter 2 not so much as how to write parallel code, as to explaining what the general factors are that can rob a parallel program of speed. This is a crucial chapter, referred to constantly in the succeeding chapters. Indeed, one can regard the entire book as addressing the plight of the poor guy described at the beginning of Chapter 2:

[1]Ironically, I myself am not a big fan of the term *data science*, but it does encompass these various views, and highlight the point that this book is about data, not physics.

> Here is an all-too-common scenario: An analyst acquires a brand new multicore machine, capable of wondrous things. With great excitement, he codes up his favorite large problem on the new machine—only to find that the parallel version runs more slowly than the serial one. What a disappointment! Let's see what factors can lead to such a situation...

The use of the word *computing* in the book's title reflects the fact that the book's main focus is indeed *computation.* This is in contrast to parallel data processing, such as in distributed file storage exemplified by that of Hadoop, though a chapter is devoted to such settings.

The main types of computing platforms covered are multicore, cluster and GPU. In addition, there is considerable coverage of Thrust, a wonderful tool that greatly eases the programming of multicore machines and GPUs—and simultaneously, in the sense that the same code is usable on either platform! I believe readers will find this material especially valuable.

One thing this book is *not*, is a user manual. Though it uses specific tools throughout, such as R's **parallel** and **Rmpi** packages, OpenMP, CUDA and so on, this is for the sake of concreteness. The book will give the reader a solid introduction to these tools, but is not a compendium of all the different function arguments, environment options and so on. The intent is that the reader, upon completing this book, will be well-poised to learn more about these tools, and most importantly, to write effective parallel code in various other languages, be it Python, Julia or whatever.

Necessary background:

If you consider yourself reasonably adept in using R, you should find most of this book quite accessible. A few sections do use C/C++, and prior background in those languages is needed if you wish to read those sections in full detail. However, even without knowing C/C++ well. you should still find that material fairly readable, and of considerable value. Appendices summarizing R for C programmers, and introducing C to R people, are included.

You should be familiar with basic math operations with matrices, mainly multiplication and addition. Occasionally some more advanced operations will be used, such as inversion (and its cousins, such as QR methods) and diagonalization, which are presented in Appendix A.

Machines:

Except when stated otherwise, all timing examples in this book were run on a 16-core Ubuntu machine, with hyperthreading degree 2. I generally used 2 to 24 cores, a range that should be similar to the platforms most readers will have available. I anticipate that the typical reader will have access to a multicore system with 4 to 16 cores, or a cluster with dozens of nodes. But even if you only have a single dual-core machine, you should still find the material here to be valuable.

For those rare and lucky readers who have access to a system consisting of thousands of cores, the material still applies, subject to the book's point that for such systems, the answer to the famous question, "Does it scale?" is often No.

CRAN packages and code:

This book makes use of several of my packages on CRAN, the R contributed software repository (`http://cran.r-project.org`): **Rdsm**, **partools** and **matpow**.

Code for the examples in this book is available at the author's Web page for the book, `http://heather.cs.ucdavis.edu/pardatasci.html`.

A note on fonts:

Function and variable names are in bold face, as are R packages (but not other packages). Code listings use the **lstlisting** package for LaTeX, usually tailored to the language used, e.g. R or C. Math italic font is used for mathematical quantities.

Thanks:

I wish to thank everyone who provided information useful to this project, either directly or indirectly. An alphabetic list would include JJ Allaire, Stuart Ambler, Matt Butner, Federico De Giuli, Matt Dowle, Dirk Eddelbuettel, David Giles, Stuart Hansen, Richard Heiberger, Bill Hsu, Michael Kane, Sameer Khan, Bryan Lewis, Mikel McDaniel, Richard Minner, George Ostrouchov, Drew Schmidt, Lars Seeman, Marc Sosnick, and Johan Wikström. I'm also very grateful to Professor Hsu for his making available to me an advanced GPU-equipped machine, and to Professor Hao Chen for use of his multicore system.

I first started developing my **Rdsm** package, used in parts of this book, at the same time that Michael Kane and Jay Emerson were developing their **bigmemory** package. Once we discovered each other's work, we had e-mail exchanges that I found quite valuable. In fact, in my later version, I decided to use **bigmemory** as a base, among other things because of its ability to use backing store. In the latter, I benefitted much from conversations with Michael, as well as with Bryan Lewis.

Much gratitude goes to the internal reviewers, David Giles, Mike Hannon and Michael Kane. I am especially grateful to my old friend Mike Hannon, who provided amazingly detailed feedback. Thanks go also to John Kimmel, Executive Editor for Statistics at Chapman and Hall, who has been highly supportive since the beginning.

My wife Gamis and my daughter Laura both have a contagious sense of humor and zest for life that greatly improve everything I do.

Author's Biography

Dr. Matloff was born in Los Angeles, and grew up in East Los Angeles and the San Gabriel Valley. He has a PhD in pure mathematics from UCLA, specializing in probability theory and statistics. He has published numerous papers in computer science and statistics, with current research interests in parallel processing, statistical computing, and regression methodology. He is on the editorial board of the *Journal of Statistical Software.*

Professor Matloff is a former appointed member of IFIP Working Group 11.3, an international committee concerned with database software security, established under UNESCO. He was a founding member of the UC Davis Department of Statistics, and participated in the formation of the UCD Computer Science Department as well. He is a recipient of the campus-wide Distinguished Teaching Award and Distinguished Public Service Award at UC Davis.

Chapter 1

Introduction to Parallel Processing in R

Instead of starting with an abstract overview of parallel programming, we'll get right to work with a concrete example in R. The abstract overview can wait. But we should place R in proper context first.

1.1 Recurring Theme: The Principle of Pretty Good Parallelism

Most of this book's examples involve the R programming language, an interpreted language. R's core operations tend to have very efficient internal implementation, and thus the language generally can offer good performance if used properly.

1.1.1 Fast *Enough*

In settings in which you really need to maximize execution speed, you may wish to resort to writing in a compiled language such as C/C++, which we will indeed do occasionally in this book. However, *the extra speed that may be attained via the compiled language typically is just not worth the effort.* In other words, we have an analogy to the Pretty Good Privacy security system:

The Principle of Pretty Good Parallelism:

In many cases just "pretty fast" is quite good enough. The extra speed we might attain by moving from R to C/C++ does not justify the possibly much longer time needed to write, debug and maintain code at that level.

This of course is the reason for the popularity of the various parallel R packages. They fulfill a desire to code parallel operations yet still stay in R. For example, the **Rmpi** package provides an R connection to the Message Passing Interface (MPI), a very widely used parallel processing system in which applications are normally written in C/C++ or FORTRAN.[1] **Rmpi** gives analysts the opportunity to take advantage of MPI while staying within R. But as an alternative to **Rmpi** that also uses MPI, R users could write their application code in C/C++, calling MPI functions, and then interface R to the resulting C /C++function. But in doing so, they would be foregoing the coding convenience and rich package available in R. So, most opt for using MPI only via the **Rmpi** interface, not directly in C/C++.

The aim of this book is to provide a general treatment of parallel processing in data science. The fact that R provides a rich set of powerful, high-level data and statistical operations means that examples in R will be shorter and simpler than they would typically be in other languages. This enables the reader to truly focus on the parallel computation methods themselves, rather than be distracted by having to wade through the details of, say, intricate nested loops. Not only is this useful from a learning point of view, but also it will make it easy to adapt the code and techniques presented here to other languages, such as Python or Julia.

1.1.2 "R+X"

Indeed, a major current trend in R is what might be called "R+X," where X is some other language or library. R+C, in which one writes one's main code in R but writes a portion needing extra speed in C or C++, has been common since the beginnings of R. These days X might be Python, Julia, Hadoop, H_2O or lots of other things.

[1] For brevity, I'll usually not mention FORTRAN, as it is not used as much in data science.

1.2 A Note on Machines

Three types of machines will be used for illustration in this book: multicore systems, clusters and graphics processing units (GPUs). As noted in the preface, I am not targeting the book to those fortunate few who have access to supercomputers (though the methods presented here do apply to such machines). Instead, it is assumed that most readers will have access to more modest systems, say multicore with 4-16 cores, or clusters with nodes numbering in the dozens, or a single GPU that may not be the absolute latest model.

Most of the multicore examples in this book were run on a 32-core system on which I seldom used all the cores (as I was a guest user). The timing experiments usually start with a small number of cores, say 2 or 4.

As to clusters, my coverage of "message-passing" software was typically run on the multicore system, though occasionally on a real cluster to demonstrate the effects of overhead.

The GPU examples here were typically run on modest hardware.

Again, the same methods as used here do apply to the more formidable systems, such as the behemoth supercomputers with multiple GPUs and so on. Tweaking is typically needed for such systems, but this is beyond the scope of this book.

1.3 Recurring Theme: Hedging One's Bets

As I write this in 2014, we are in an exciting time for parallel processing. Hardware can be purchased for the home that has speeds and degrees of parallelism that were unimaginable for ordinary PC users just a decade ago. There have been tremendous improvements in software as well; R now has a number of approaches to parallelism, as we will see in this book; Python finally threw off its GIL yoke a few years ago;[2] C++11 has built-in parallelism; and so on.

Due to all this commotion, though, we really are in a rather unsettled state as we look forward to the coming years. Will GPUs become more and more mainstream? Will multicore chips routinely contain so many cores that GPUs remain a nice type of hardware? Will accelerator chips like the

[2]This is the Global Interpreter Lock, which prevented true parallel operation in Python. The GIL is still there, but Python now has ways around it.

Intel Xeon Phi overtake GPUs? And will languages keep pace with the advances in hardware?

For this reason, software packages that span more than one type of hardware, and can be used from more than one type of programming language, have great appeal. The Thrust package, the topic of Chapter 7 and some of the later sections, epitomizes this notion. The same Thrust code can run on either multicore or GPU platforms, and since it is C++ based, it is accessible from R or most other languages. In short, Thrust allows us to "hedge our bets" when we develop parallel code.

Message-passing software systems, such as R's **snow**, **Rmpi** and **pbdR**, have much the same advantage, as they can run on either multicore machines or clusters.

1.4 Extended Example: Mutual Web Outlinks

So, let's look at our promised concrete example.

Suppose we are analyzing Web traffic, and one of our questions concerns how often two websites have links to the same third site. Say we have outlink information for n Web pages. We wish to find the mean number of mutual outlinks per pair of sites, among all pairs.

This computation is actually similar in pattern to those of many statistical methods, such as Kendall's τ and the U-statistic family. The pattern takes the following form. For data consisting of n observations, the pattern is to compute some quantity g for each pair of observations, then sum all those values, as in this pseudocode (i.e., outline):

```
sum = 0.0
for i = 1,...,n-1
   for j = i+1,,...,n
      sum = sum + g(obs.i, obs.j)
```

With nested loops like this, you'll find in this book that it is generally easier to parallelize the outer loop rather than the inner one. If we have a dual core machine, for instance, we could assign one core to handle some values of **i** in the above code and the other core to handle the rest. Ultimately we'll do that here, but let's first take a step back and think about this setting.

1.4.1 Serial Code

Let's first implement this procedure in serial code:

```
mutoutser <- function(links) {
   nr <- nrow(links)
   nc <- ncol(links)
   tot <- 0
   for (i in 1:(nr-1)) {
      for (j in (i+1):nr) {
         for (k in 1:nc)
            tot <- tot + links[i,k] * links[j,k]
      }
   }
   tot / (nr * (nr-1) / 2)
}
```

Here **links** is a matrix representing outlinks of the various sites, with **links[i,j]** being 1 or 0, according to whether there is an outlink from site **i** to site **j**. The code is a straightforward implementation of the pseudocode in Listing 1.4.1 above.

How does this code do in terms of performance? Consider this simulation:

```
sim <- function(nr,nc) {
   lnk <- matrix(sample(0:1,(nr*nc),replace=TRUE),
     nrow=nr)
   system.time(mutoutser(lnk))
}
```

We generate random 1s and 0s, and call the function. Here's a sample run:

```
> sim(500,500)
   user  system elapsed
106.111   0.030 106.659
```

Elapsed time of 106.659 seconds—awful! We're dealing with 500 websites, a tiny number in view of the millions that are out there, and yet it took almost 2 minutes to find the mean mutual outlink value for this small group of sites.

It is well known, though, that explicit **for** loops are slow in R, and here we have two of them. The first solution to try for loop avoidance is *vectorization*, meaning to replace a loop with some vector computation. This gives one the speed of the C code that underlies the vector operation, rather than having to translate the R repeatedly for each line of the loop, at each iteration.

In the code for **mutoutser()** above, the inner loops can be rewritten as a matrix product, as we will see below, and that will turn out to eliminate two of our loops.[3]

To see the matrix formulation, suppose we have this matrix:

$$\begin{pmatrix} 0 & 1 & 0 & 0 & 1 \\ 1 & 0 & 0 & 1 & 1 \\ 0 & 1 & 0 & 1 & 0 \\ 1 & 1 & 1 & 0 & 0 \\ 1 & 1 & 1 & 0 & 1 \end{pmatrix} \tag{1.1}$$

Consider the case in which **i** is 2 and **j** is 4 in the above pseudocode, Listing 1.4.1. The innermost loop, i.e., the one involving **k**, computes

$$1 \cdot 1 + 0 \cdot 1 + 0 \cdot 1 + 1 \cdot 0 + 1 \cdot 0 = 1 \tag{1.2}$$

But that is merely the inner product of rows **i** and **j** of the matrix! In other words, it's

```
links[i,] %*% links[j,]
```

But there's more. Again consider the case in which **i** is 2. The same reasoning as above shows that the entire computation for all **j** and **k**, i.e., the two innermost loops, can be written as

$$\begin{pmatrix} 0 & 1 & 0 & 1 & 0 \\ 1 & 1 & 1 & 0 & 0 \\ 1 & 1 & 1 & 0 & 1 \end{pmatrix} \begin{pmatrix} 1 \\ 0 \\ 0 \\ 1 \\ 1 \end{pmatrix} = \begin{pmatrix} 1 \\ 1 \\ 2 \end{pmatrix} \tag{1.3}$$

The matrix on the left is the portion of our original matrix below row 2, and the vector on the right is row 2 itself.

Those numbers, 1, 1 and 2, are the results we would get from running the code with **i** equal to 2 and **j** equal to 3, 4 and 5. (Check this yourself to get a better grasp of how this works.)

[3]In R, a matrix is a special case of a vector, so we are indeed using vectorization here, as promised.

So, we can eliminate two loops, as follows:

```
mutoutser1<- function(links) {
   nr <- nrow(links)
   nc <- ncol(links)
   tot <- 0
   for (i in 1:(nr-1)) {
      tmp <- links[(i+1):nr,] %*% links[i,]
      tot <- tot + sum(tmp)
   }
   tot / nr
}
```

This actually brings a dramatic improvement:

```
sim1 <- function(nr,nc) {
   lnk <- matrix(sample(0:1,(nr*nc),replace=TRUE),
      nrow=nr)
   print(system.time(mutoutser1(lnk)))
}

> sim1(500,500)
   user  system elapsed
  1.443   0.044   1.496
```

Wonderful! Nevertheless, that is still only for the very small 500-site case. Let's run it for 2000:

```
> sim1(2000,2000)
   user  system elapsed
 92.378   1.002  94.071
```

Over 1.5 minutes! And 2000 is still not very large.

We could further fine-tune our code, but it does seem that parallelizing may be a better option. Let's go that route.

1.4.2 Choice of Parallel Tool

The most popular tools for parallel R are **snow**, **multicore**, **foreach** and **Rmpi**. Since the first two of these are now part of the R core in a package named **parallel**, it is easiest to use one of them for our introductory material in this chapter, rather than having the user install another package at this point.

Our set of choices is further narrowed by the fact that **multicore** runs only on Unix-family (e.g., Linux and Mac) platforms, not Windows. Accordingly, at this early point in the book, we will focus on **snow**.

1.4.3 Meaning of "snow" in This Book

As noted, an old contributed package for R, **snow**, was later made part of the R base, in the latter's **parallel** package (with slight modifications). We will make frequent use of this part of that package, so we need a short name for it. "The portion of **parallel** adapted from **snow**" would be anything but short. So, we'll just call it **snow**.

1.4.4 Introduction to snow

Here is the overview of how **snow** operates: All four of the popular packages cited above, including **snow**, typically employ a *scatter/gather* paradigm: We have multiple instances of R running at the same time, either on several machines in a cluster, or on a multicore machine. We'll refer to one of the instances as the *manager*, with the rest being *workers*. The parallel computation then proceeds as follows:

- **scatter:** The manager breaks the desired computation into chunks, and sends ("scatters") the chunks to the workers.
- **chunk computation**: The workers then do computation on each chunk, and send the results back to the manager.
- **gather:** The manager receives ("gathers") those results, and combines them to solve the original problem.

In our mutual-outlink example here, each chunk would consist of some values of **i** in the outer **for** loop in Listing 1.4.1. In other words, each worker would determine the total count of mutual outlinks for this worker's assigned values of **i**, and then return that count to the manager. The latter would collect these counts, sum them to form the grand total, and then obtain the average by dividing by the number of node pairs, n(n-1)/2.

1.4.5 Mutual Outlinks Problem, Solution 1

Here's our first cut at the mutual outlinks problem:

1.4.5.1 Code

```
library(parallel)

doichunk <- function(ichunk) {
   tot <- 0
   nr <- nrow(lnks)  # lnks global at worker
   for (i in ichunk) {
      tmp <- lnks[(i+1):nr,] %*% lnks[i,]
      tot <- tot + sum(tmp)
   }
   tot
}

mutoutpar <- function(cls,lnks) {
   nr <- nrow(lnks)  # lnks global at manager
   clusterExport(cls,"lnks")
   # each "chunk" has only 1 value of i, for now
   ichunks <- 1:(nr-1)
   tots <- clusterApply(cls,ichunks,doichunk)
   Reduce(sum,tots) / nr
}

snowsim <- function(nr,nc,cls) {
   lnks <<-
      matrix(sample(0:1,(nr*nc),replace=TRUE),
         nrow=nr)
   system.time(mutoutpar(cls,lnks))
}

# set up cluster of nworkers workers on
# multicore machine
initmc <- function(nworkers) {
   makeCluster(nworkers)
}

# set up a cluster on machines specified,
# one worker per machine
initcls <- function(workers) {
   makeCluster(spec=workers)
}
```

1.4.5.2 Timings

Before explaining how this code works, let's see if it yields a speed improvement. I ran on the same machine used earlier, but in this case with two workers, i.e., on two cores. Here are the results:

```
> cl2 <- initmc(2)
> snowsim(2000,2000,cl2)
   user  system elapsed
  0.237   0.047  80.348
```

So we did get a speedup, with run time being diminished by almost 14 seconds. Good, but note that the speedup factor is only 94.071/80.348 = 1.17, not the 2.00 one might expect from using two workers. This illustrates that communication and other overhead can indeed be a major factor.

Note the stark discrepancy between `user` and `elapsed` time here. Remember, these are times for the manager! The main computation is done by the workers, and their times don't show up here except in `elapsed` time.

You might wonder whether two cores are enough, since we have a total of three processes—two workers and the manager. But since the manager is idle while the two workers are computing, there would be no benefit in having the manager run on a separate core, even if we had one (which we in a sense do, with hyperthreading, to be explained shortly).

This run was performed on a dual core machine, hence our using two workers. However, we may be able to do a bit better, as this machine has a *hyperthreaded* processor. This means that each core is capable, to some degree, of running two programs at once. Thus I tried running with four workers:

```
> cl2 <- initmc(4)
> snowsim(2000,2000,cl2)
   user  system elapsed
  0.484   0.051  70.077
```

So, hyperthreading did yield further improvement, raising our speedup factor to 1.34. Note, though, that now there is even further disparity between the 4.00 speedup we might hope to get with four workers. As noted, these issues will arise frequently in this book; the sources of overhead will be discussed, and remedies presented.

There is another reason why our speedups above are not so impressive: Our code is fundamentally unfair—it makes some workers do more work than others. This is known as a *load balancing* problem, one of the central issues in the parallel processing field. We'll address this in a refined version in Chapter 3.

1.4.5.3 Analysis of the Code

So, how does all this work? Let's dissect the code.

Even though **snow** and **multicore** are now part of R via the **parallel** package, the package is not automatically loaded. So we need to take care of this first, by placing a line

```
library(parallel)
```

at the top of our source file (if all these functions are in one file), or simply execute the above **library()** call on the command line.

Or, we can insert a line

```
require(parallel)
```

in the functions that make use of **snow**.

Now, who does what? It's important to understand that most of the lines of code in the serial version are executed by the manager. The only code run by the workers will be **doichunk()**, though of course that is where the main work is done. As will be seen, the manager sends that function (and data) to the workers, who execute the function according to the manager's directions.

The basic idea is to break the values of **i** in the **i** loop in our earlier serial code, Listing 1.4.1, into chunks, and then have each worker work on its chunk. Our function **doichunk()** ("do element i in ichunk"),

```
doichunk <- function(ichunk) {
   tot <- 0
   nr <- nrow(lnks)  # lnks global at worker
   for (i in ichunk) {
      tmp <- lnks[(i+1):nr,] %*% lnks[i,]
      tot <- tot + sum(tmp)
   }
   tot
}
```

will be executed for each worker, with **ichunk** being different for each worker.

Our function **mutoutpar()** wraps the overall process, dividing into the **i** values into chunks and calling **doichunk()** on each one. It thus parallelizes the outer loop of the serial code.

```
mutoutpar <- function(cls,lnks) {
   nr <- nrow(lnks)
   clusterExport(cls,"lnks")
   ichunks <- 1:(nr-1)
   tots <- clusterApply(cls,ichunks,doichunk)
   Reduce(sum,tots) / nr
}
```

To get an overview of that function, note that the main actions consist of the follwing calls to **snow** and R functions:

- We call **snow**'s **clusterExport()** to send our data, in this case the **lnks** matrix, to the workers.
- We call **snow**'s **clusterApply()** to direct the workers to perform their assigned chunks of work.
- We call R's core function **Reduce()** as a convenient way to combine the results returned by the workers.

Here are the details: Even before calling **mutoutpar()**, we set up our **snow** cluster:

```
makeCluster(nworkers)
```

This sets up **nworkers** workers. Remember, each of these workers will be separate R processes (as will be the manager). In this simple form, they will all be running on the same machine, presumably multicore.

Clusters are **snow** abstractions, not physical entities, though we can set up a **snow** cluster on a physical cluster of machines. As will be seen in detail later, a cluster is an R object that contains information on the various workers and how to reach them. So, if I run

```
cls <- initmc(4)
```

I create a 4-node **snow** cluster (for 4 workers) and save its information in an R object **cls** (of class **"cluster"**), which will be used in my subsequent calls to **snow** functions.

There is one component in **cls** for each worker. So after the above call, running

```
length(cls)
```

prints out 4.

We can also run **snow** on a physical cluster of machines, i.e., several machines connected via a network. Calling the above function **initcls()** arranges this. In my department, for example, we have student lab machines named **pc1**, **pc2** and so on, so for instance

```
cl2 <- initcls(c("pc28","pc29"))
```

would set up a two-node **snow** run.

In any case, in the above default call to **makeCluster()**, communication between the manager and the workers is done via network sockets, even if we are on a multicore machine.

Now, let's take a closer look at **mutoutpar()**, first the call

```
clusterExport(cls,"lnks")
```

This sends our data matrix **lnks** to all the workers in **cls**.

An important point to note is that **clusterExport()** by default requires the transmitted data to be global in the manager's work space. It is then placed in the global work space of each worker (without any alternative option offered). To meet this requirement, I made **lnks** global back when I created this data in **snowsim()**, using the superassignment operator <<−:

```
lnks <<- matrix(sample(0:1,(nr*nc),replace=TRUE),
   nrow=nr)
```

The use of global variables is rather controversial in the software development world. In my book *The Art of R Programming* (NSP, 2011), I address some of the objections some programmers have to global variables, and argue that in many cases (especially in R), globals are the best (or least bad) solution.

In any case, here the structure of **clusterExport()** basically forces us to use globals. For the finicky, there is an option to use an R environment instead of the manager's global workspace. We could change the above call with **mutoutpar()**, for instance, to

```
clusterExport(cls,"lnks",envir=environment())
```

The R function **environment()** returns the current environment, meaning the context of code within **mutoutpar()**, in which **lnks** is a local variable. But even then the data would still be global at the workers.

Here are the details of the **clusterApply()** call. Let's refer to that second argument of **clusterApply()**, in this case **ichunks**, as the "work assignment" argument, as it parcels out work to workers.

To keep things simple in this introductory example, we have just a single **i** value for each "chunk":

```
ichunks <- 1:(nr-1)
tots <- clusterApply(cls,ichunks,doichunk)
```

(We'll extend this to larger chunks in Section 3.2.1.)

Here **clusterApply()** will treat that **ichunks** vector as an R list of **nr - 1** elements. In the call to that function, we have the manager sending **ichunks**[[**1**]] to **cls**[[**1**]], which is the first worker. Similarly, **ichunks**[[**2**]] is sent to **cls**[[**2**]], the second worker, and so on.

Unless the problem is small (far too small to parallelize!), we will have more chunks than workers here. The **clusterApply()** function handles this in a *Round Robin* manner. Say we have 1000 chunks and 4 workers. After **clusterApply()** sends the fourth chunk to the fourth worker, it starts over again, sending the fifth chunk to the first worker, the sixth chunk to the second worker, and so on, repeatedly cycling through the workers. In fact, the internal code uses R's *recycling* operation to implement this.

Each worker is told to run **doichunk()** on each chunk sent to that worker by the manager. The second worker, for example, will call **doichunk()** on **ichunks**[[**2**]], **ichunks**[[**6**]], etc.

So, each worker works on its assigned chunks, and returns the results—the number of mutual outlinks discovered in the chunks—to the manager. The **clusterApply()** function collects these results, and places them into an R list. which we've assigned here to **tots**. That list will contain **nr - 1** elements.

One might expect that we could then find the grand sum of all those totals returned by the workers by simply calling R's **sum()** function:

```
sum(tots)
```

This would have been fine if **tots** had been a vector, but it's a list, hence our use of R's **Reduce()** function. Here **Reduce()** will apply the **sum()** function to each element of the list **tots**, yielding the grand sum as desired. You'll find use of **Reduce()** common with functions in packages like **snow**, which typically return values in lists.

This is a good time to point out that *many parallel R packages require the user to be adept at using R lists.* Our call to **clusterApply()**, returned a list type, and in fact its second argument is usually an R list, though not here.

This example has illustrated some of the major issues, but it has barely scratched the surface. The next chapter will begin to delve deeper into this many-faceted subject.

1.5 Further Reading

As noted, the R list data type is central to parallel computing in R. There is some material on this in Appendix B, with further details in my book, *The Art of R Programming*, NSP, 2011, and in numerous online R tutorials.

Chapter 2

"Why Is My Program So Slow?": Obstacles to Speed

Here is an all-too-common scenario: An analyst acquires a brand new multicore machine, capable of wondrous things. With great excitement, he codes up his favorite large problem on the new machine—only to find that the parallel version runs more slowly than the serial one. What a disappointment!

Though you are no doubt eager to get to some more code, a firm grounding in the infrastructural issues will prove to be quite valuable indeed, hence the need for this chapter. These issues will arise repeatedly in the rest of the book. If you wish, you could skip ahead to the other chapters now, and come back to this one as the need arises, but it's better if you go through it now. So, let's see what factors can lead to such a situation in which our hapless analyst above sees his wonderful plans go awry.

2.1 Obstacles to Speed

Let's refer to the computational entities as *processes*, such as the workers in the case of **snow**. There are two main performance issues in parallel programming:

- *Communications overhead:* Typically data must be transferred back and forth between processes. This takes time, which can take quite a toll on performance.

  In addition, the processes can get in each other's way if they all try to access the same data at once. They can collide when trying to access the same communications channel, the same memory module, and so on. This is another sap on speed.

  The term *granularity* is used to refer, roughly, to the ratio of computation to overhead. *Large-grained* or *coarse-grained* algorithms involve large enough chunks of computation that the overhead isn't much of a problem. In *fine-grained* algorithms, we really need to avoid overhead as much as possible.

- *Load balance:* As noted in the last chapter, if we are not careful in the way in which we assign work to processes, we risk assigning much more work to some than to others. This compromises performance, as it leaves some processes unproductive at the end of the run, while there is still work to be done.

There are a number of issues of this sort that occur generally enough to be collected into this chapter, as an "early warning" of issues that can arise. This is just an overview, with details coming in subsequent chapters, but being forewarned of the problems will make it easier to recognize them as they are encountered.

2.2 Performance and Hardware Structures

Scorecards, scorecards! You can't tell the players without the scorecards!—old chant of scorecard vendors at baseball games

The foot bone connected to the ankle bone, The ankle bone connected to the shin bone...—from the children's song, "Dem Bones"

The reason our unfortunate analyst in the preceding section was surprised that his code ran more slowly on the parallel machine was almost certainly due to a lack of understanding of the underlying hardware and systems software. While one certainly need not understand the hardware on an electronics level, a basic knowledge of "what is connected to what" is essential.

In this section, we'll present overviews of the major hardware issues, and of the two parallel hardware technologies the reader is mostly likely to encounter, *multiprocessors* and *clusters*:[1]

- A multiprocessor system has, as the name implies, two or more processors, i.e., two or more CPUs, so that two or more programs (or parts of the same program) can be doing computation at the same time. A *multicore* system, common in the home, is essentially a low-end multiprocessor, as we will see later. Multiprocessors are also known as *shared-memory* systems, since they indeed share the same physical RAM.

  These days, almost any home PC or laptop is at least dual core. If you own such a machine, congratulations, you own a multiprocessor system!

  You are also to be congratulated for owning a multiprocessor system if you have a fairly sophisticated video card in your computer, one that can serve as a *graphics processing unit.* GPUs are specialized shared-memory systems.

- A cluster consists of multiple computers, each capable of running independently, that are networked together, enabling their engaging in a concerted effort to solve a big numerical problem.

  If you have a network at home, say with a wireless or wired router, then congratulations, you own a cluster![2]

I emphasize the "household item" aspect above, to stress that these are not esoteric architectures, though of course scale can vary widely from what you have at home to far more sophisticated and expensive systems, with quite a bit in between.

The terms *shared-memory* and *networked* above give clues as to the obstacles to computational speed that arise, which are key. So, we will first discuss the high-level workings of these two hardware structures, in Sections 2.3 and 2.4.

We'll then explain how they apply to the overhead issue with our two basic platform types, multicore (Section 2.5.1.1) and cluster (Section 2.5.1.2). We'll cover just enough details to illustrate the performance issues discussed later in this chapter, and return for further details in later chapters.

---

[1]What about clouds? A cloud consists of multicore machines and clusters too, but operating behind the scenes.

[2]It should be noted that in the case of large clusters used for intensive computation, one generally must install software for the purpose of controlling which program runs on which machines. But your two-node home system is still a cluster.

2.3 Memory Basics

Slowness of memory access is one of the most common issues arising in high-performance computing. Thus a basic understanding of memory is vital.

Consider an ordinary assignment statement, copying one variable (a single integer, say) to another:

```
y = x
```

Typically, both **x** and **y** will be stored somewhere in memory, i.e., RAM (Random Access Memory). Memory is broken down into *bytes*, designed to hold one character, and *words*, usually designed to contain one number. A byte consists of eight bits, i.e., eight 0s and 1s. On typical computers today, the word size is 64 bits, or eight bytes.

Each word has an ID number, called an *address*. (Individual bytes have addresses too, but this will not concern us here.) So the compiler (in the case of C/C++/FORTRAN) or the interpreter (in the case of a language like R), will assign specific addresses in memory at which **x** and **y** are to be stored. The above assignment will be executed by the machine's copying one word to the other.

A vector will typically be stored in a set of consecutive words. This will be the case for matrices too, but there is a question as to whether this storage will be row-by-row or column-by-column. C/C++ uses *row-major order*: First all of the first row (called row 0) is stored, then all of the second row, and so on. R and FORTRAN use *column-major order*, storing all of the first column (named column 1) etc. So, for instance, if **z** is a 5×8 matrix in R, then **z[2,3]** will be in the 12th word (5+5+2) in the portion of memory occupied by **z**. These considerations will affect performance, as we will see later.

Memory access time, even though measured in tens of nanoseconds—billionths of a second—is slow relative to CPU speeds. This is due not only to electronic delays within the memory chips themselves, but also due to the fact that the pathway to memory is often a bottleneck. More on this below.

2.3.1 Caches

A device commonly used to deal with slow memory access is a *cache*. This is a small but fast chunk of memory that is located on or near the processor chip. For this purpose, memory is divided into *blocks*, say of 64 bytes each. Memory address 1200, for instance, would be in block 18, since 1200/64 is equal to 18 plus a fraction. (The first block is called Block 0.)

The cache is divided into *lines*, each the size of a memory block. At any given time, the cache contains local copies of some blocks of memory, with the specific choice of blocks being dynamic—at some times the cache will contain copies of some memory blocks, while a bit later it may contain copies of some other blocks.[3]

If we are lucky, in most cases, the memory word that the processor wishes to access (i.e., the variable in the programmer's code she wishes to access) already has a copy in its cache—a *cache hit.* If this is a read access (of **x** in our little example above), then it's great—we avoid the slow memory access.

On the other hand, in the case of a write access (to **y** above), if the requested word is currently in the cache, that's nice too, as it saves us the long trip to memory (if we do not "write through" and update memory right away, as we are assuming here). But it does produce a discrepancy between the given word in memory and its copy in the cache. In the cache architecture we are discussing here, that discrepancy is tolerated, and eventually resolved when the block in question is "evicted," as we will see below. (With a multicore machine, cache operation becomes more complicated, as typically each core will have its own cache, thus potentially causing severe discrepancies. This will be discussed in Section 2.5.1.1.)

If in a read or write access the desired memory word is not currently in the cache, this is termed a *cache miss.* This is fairly expensive. When it occurs, the entire block containing the requested word must be brought into the cache. In other words, we must access many words of memory, not just one. Moreover, usually a block currently in the cache must be *evicted* to make room for the new one being brought in. If the old block had been written to at all, we must now write that entire block back to memory, to update the latter.[4]

So, though we save memory access time when we have a cache hit, we incur a substantial penalty at a miss. Good cache design can make it so that the penalty is incurred only rarely. When a read miss occurs, the hardware makes "educated guesses" as to which blocks are least likely to be needed again in the near future, and evicts one of these. It usually guesses well, so

[3]What follows below is a description of a common cache design. There are many variations, not discussed here.

[4]There is a *dirty bit* that records whether we've written to the block, but not which particular words were affected. Thus the entire block must be written.

that cache hit rates are typically well above 90%. Note carefully, though, that this can be affected by the way we code. This will be discussed in future chapters.

A machine will typically have two or more levels of cache. The one in or next to the CPU is called the L1, or Level 1 cache. Then there may be an L2 cache, a "cache for the cache." If the desired item is not found in the L1 cache, the CPU will then search the L2 cache before resorting to accessing the item in memory.

2.3.2 Virtual Memory

Though it won't arise much in our context, we should at least briefly discuss *virtual memory.* Consider our example above, in which our program contained variables **x** and **y**. Say these are assigned to addresses 200 and 8888, respectively. Fine, but what if another program is also running on the machine? The compiler/interpreter may have assigned one of its variables, say **g**, to address 200. How do we resolve this?

The standard solution is to make the address 200 (and all others) only "virtual." It may be, for instance, that **x** from the first program is actually stored in physical address 7260. The program will still say **x** is at word 200, but the hardware will translate 200 to 7260 as the program executes. If **g** in the second program is actually in word 6548, the hardware will replace 200 by 6548 every time the program requests access to word 200. The hardware has a table to do these lookups, one table for each program currently running on the machine, with the table being maintained by the operating system.

Virtual memory systems break memory into *pages*, say of 4096 bytes each, analogous to cache blocks. Usually, only some of your program's pages are *resident* in memory at any given time, with the remainder of the pages out on disk. If your program needs some memory word not currently resident—a *page fault*, analogous to a cache miss—the hardware senses this, and transfers control to the operating system. The OS must bring in the requested page from disk, an extremely expensive operation in terms of time, due to the fact that a disk drive is mechanical rather than electronic like RAM.[5] Thus page faults can really slow down program speed, and again as with the cache case, you may be able to reduce page faults through careful design of your code.

[5]Some more expensive drives, known as Solid State Drives (SSDs), are in fact electronic.

2.3.3 Monitoring Cache Misses and Page Faults

Both cache misses and page faults are enemies of good performance, so it would be nice to monitor them.

This actually can be done in the case of page faults. As noted, a page fault triggers a jump to the OS, which can thus record it. In Unix-family systems, the **time** command gives not only run time but also a count of page faults.

By contrast, cache misses are handled purely in hardware, thus not recordable by the OS. But one might try to gauge the cache behavior of a program by using the number of page faults as a proxy. There are also simulators, such as **valgrind**, which can be used to measure cache performance.

2.3.4 Locality of Reference

Clearly, the effectiveness of caches and virtual memory depend on repeatedly using items in the same blocks (*spatial locality*) within short time periods (*temporal locality*). As mentioned earlier, this in turn can be affected to some degree by the way the programmer codes things.

Say we wish to find the sum of all elements in a matrix. Should our code traverse the matrix row-by-row or column-by-column? In R, for instance, which as mentioned stores matrices in column-major order, we should go column-by-column, to get better locality.

A detailed case study on cache behavior will be presented in Section 5.8.

2.4 Network Basics

A single Ethernet (or other similar system), say within a building, is called a *network*. The *Internet* is simply the interconnection of many networks–millions of them.

Say you direct the browser on your computer to go to the Cable Network News (CNN) home page, and you are located in San Francisco. Since CNN is headquartered in Atlanta, *packets* of information will go from San Francisco to Atlanta. (Actually, they may not go that far, since Internet service providers (ISPs) often cache Web pages, but let's suppose that doesn't occur.) Actually, a packet's journey will be rather complicated:

- Your browser program will write your Web request to a *socket.* The latter is not a physical object, but rather a software interface from your program to the network.

- The socket software will form a packet from your request, which will then go through several layers of the *network protocol stack* in your OS. Along the way, the packet will grow, as more information is being added, but also it will split into multiple, smaller packets.

- Eventually the packets will reach your computer's network interface hardware, from which they go onto the network.

- A *gateway* on the network will notice that the ultimate destination is external to this network, so the packets will be transferred to another network that the gateway is also attached to.

- Your packets will wend their way across the country, being sent from one network to the next.[6]

- When your packets reach a CNN computer, they will now work their way *up* the levels of the OS, finally reaching the Web server program.

2.5 Latency and Bandwidth

Getting there is half the fun—old saying, regarding the pleasures of traveling

The speed of a communications channel—whether between processor cores and memory in shared-memory platforms, or between network nodes in a cluster of machines—is measured in terms of *latency*, the end-to-end travel time for a single bit, and *bandwidth*, the number of bits per second that we can pump onto the channel.

To make the notions a little more concrete, consider the San Francisco Bay Bridge, a long, multilane structure for which westbound drivers pay a toll. The notion of latency would describe the time it takes for a car to drive from one end of the bridge to the other. (For simplicity, assume they all go the same speed.) By contrast, the bandwidth would be the number of cars exiting from the toll booths per unit time. We can reduce the latency by raising the speed limit on the bridge, while we could increase the bandwidth by adding more lanes and more toll booths.

---

[6]Run the **traceroute** command on your machine to see the exact path, though this can change over time.

The network time in seconds to send an n-byte message, with a latency of l seconds and a bandwidth of b bytes/second, is clearly

$$l + n/b \tag{2.1}$$

Of course, this assumes that there are no other messages contending for the communication channel.

Clearly there are numerous delays in networks, including the less-obvious ones incurred in traversing the layers of the OS. Such traversal involves copying the packet from layer to layer, and in cases of interest in this book, such copying can involve huge matrices and thus take a lot of time.

Though parallel computation is typically done within a network rather than across networks as above, many of those delays are still there. So, network speeds are much, much lower than processor speeds, both in terms of latency and bandwidth.

The latency in even a fast network such as Infiniband is on the order of microseconds, i.e., millionths of a second, which is eons compared to the nanosecond level of execution time for a machine instruction in a processor. (Beware of a network that is said to be fast but turns out only to have high bandwidth, not also low latency.)

Latency and bandwidth issues arise in shared-memory systems too. Consider GPUs, for instance. In most applications, there is a lot of data transfer between the CPU and the GPU, with attendant potential for slowdown. Latency, for example, is the time for a single bit to go from the CPU to the GPU, or vice versa.

One way to ameliorate the slowdown from long latency delays is *latency hiding*. The basic idea is to try to do other useful work while a communication having long latency is pending. This approach is used, for instance, in the use of nonblocking I/O in message-passing systems (Section 8.7.1) to deal with network latency, and in GPUs (Chapter 6) to deal with memory latency.

2.5.1 Two Representative Hardware Platforms: Multicore Machines and Clusters

Multicore machines have become standard on the desktop (even in the cell phone!), and many data scientists have access to computer clusters. What are the performance issues on these platforms? The next two sections

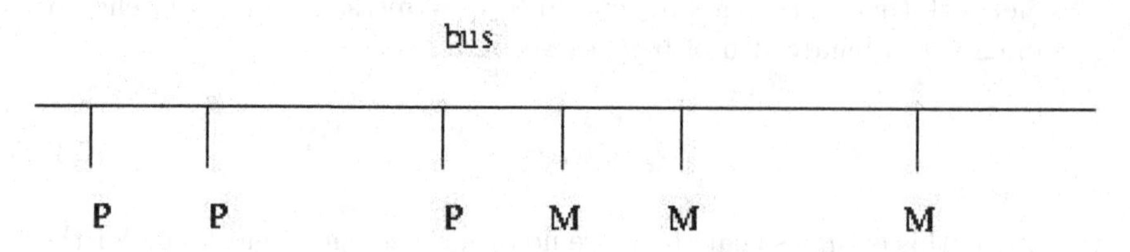

Figure 2.1: Symmetric Multiprocsssor System

provide an overview.

2.5.1.1 Multicore

A *symmetric multiprocessor system* looks something like Figure 2.1 in terms of components and, most importantly, their interconnection. What do we see?

- There are *processors*, depicted by the Ps, in which your program is physically executed.
- There are *memory banks*, the Ms, in which your program and data reside during execution.[7]
- The processors and memory banks are connected to a *bus*, a set of parallel wires used for communication between these computer components.

Your input/output hardware—disk drives, keyboards and so on—are also connected to the bus, and there may actually be more than one bus, but our focus will be mainly on the processors and memory.

A *threaded* program will have several instantiations of itself, called *threads*, that are working in concert to achieve parallelism. They run independently, except that they share the data of the program in common. If your program is threaded, it will be running on several of the processors at once, each thread on a different core. A key point, as we will see, is that the shared

[7]These were called *banks* in the old days. Later the term *modules* became more popular, but with the recent popularity of GPUs, the word *banks* has come back into favor.

memory becomes the vehicle for communication between the various processes.

Your program consists of a number of machine language instructions. (If you write in an interpreted language such as R, the interpreter itself consists of such instructions.) As the processors execute your program, they will fetch the instructions from memory.

As noted earlier, your data—the variables in your program—is stored in memory. The machine instructions fetch the data from memory as needed, so that it can be processed, e.g., summed, in the processors.

Until recently, ordinary PCs sold at your local electronics store followed the model in Figure 2.1 but with only one P. Multiprocessor systems enabled parallel computation, but cost hundreds of thousands of dollars. But then it became standard for systems to have a *multicore* form. This means that there are multiple Ps, but with the important distinction that they are all on a single chip (each P is one core), making for inexpensive systems.[8] Whether on a single chip or not, having multiple Ps sets up parallel computation, and is known as the *shared-memory* paradigm, for obvious reasons.

By the way, why are there multiple Ms in Figure 2.1? To improve memory performance, the system is set up so that memory is partitioned into several banks (typically there are the same number of Ms as Ps). This enables us to not only do *computation* on a parallel basis—several Ps working on different pieces of a problem in parallel—but also to do *memory access* in parallel—several memory accesses being active in parallel, in different banks. This amortizes the memory access penalty. Of course, if more than one P happens to need to access the same M at about the same time, we lose this parallelism.

As you can see, a potential bottleneck is the bus. When more than one P needs to access memory at a time, even if to different banks, attempting to place memory access requests on the bus, all but one of them will need to wait. This *bus contention* can cause significant slowdown. Much more elaborate systems, featuring multiple communications channels to memory rather than just a bus, have also been developed and serve to ameliorate the bottleneck issue. Most readers of this book, however, are more likely to use a multicore system on a single memory bus.

You can see now why efficient memory access is such a crucial factor in achieving high performance. There is one more tool to handle this that is

[8]Terminology is not standardized, unfortunately. It is common to refer to that chip as "the" processor, even though there actually are multiple processors inside.

vital to discuss here: Use of caches. Note the plural; in Figure 2.1, there is usually a C in between each P and the bus.

As with uniprocessor systems, caching can bring a big win in performance. In fact, the potential is even greater with a multiprocessor system, since caching will now bring the additional benefit of reducing bus contention. Unfortunately, it also produces a new problem, *cache coherency*, as follows.[9]

Consider what happens upon a *write hit*, i.e., a write to a location for which a local cache copy exists. For instance, consider code such as

```
x = 28;
```

with **x** having address 200. This code might be executed at a time when there is a copy of word 200 in that processor's cache. The problem is that other caches may also have a copy of this word, so they are now invalid for that block. (Recall that validity is defined only at the block level; if all words in a block but one are valid, the whole block is considered invalid.) The hardware must now inform those other caches that their copies of this block are invalid.

The hardware does so via the bus, thus incurring an expensive bus operation. Moreover, the next time this word (or for that matter, any word in this block) is requested at one of the other caches, there will be a cache miss, again an expensive event.

Once again, proper coding on the programmer's part can sometimes ameliorate the cache coherency problem.

A final point on multicore structure: Even on a uniprocessor machine, one generally has multiple programs running concurrently. You might have your browser busy downloading a file, say, while at the same time you are using a photo processing application. With just a single processor, these programs will actually take turns running; each one will run for a short time, say 50 milliseconds, then hand off the processor to the next program, in a cyclic manner. (You as the user probably won't be aware of this directly, but you may notice the system as a whole slowing down.) Note by the way that if a program is doing a lot of input/output (e.g., file access), it is effectively idle during I/O times; as soon as it starts an I/O operation, it will relinquish the processor.

By contrast, on a multicore machine, you can have multiple programs running physically simultaneously (though of course they will still take turns if there are more of them than there are cores).

[9]As noted earlier, there are variations of the structure described here, but this one is typical.

2.5.1.2 Clusters

These are much simpler to describe, though with equally thorny performance obstacles.

The term *cluster* simply refers to a set of independent *processing elements* (PEs) or *nodes* that are connected by a local area network, such as the common Ethernet or the high-performance Infiniband. Each PE consists of a CPU and some RAM. The PE could be a full desktop computer, including keyboard, disk drive and monitor, but if it is used primarily for parallel computation, then just one monitor, keyboard and so on suffice for the entire system. A cluster may also have a special operating system, to coordinate assigning of user programs to PEs.

We will have one computational process per PE (unless each PE is a multicore system, as is common). Communication between the processes occurs via the network. The latter aspect, of course, is where the major problems occur.

2.5.2 The Principle of "Just Leave It There"

All of these considerations regarding latency and bandwidth means, among other things, that data copying is often the enemy of speed. This is particularly true for high-latency platforms such as clusters and GPUs.

In such a situation, it may be crucial to design the algorithm to minimize copying. With an iterative algorithm, for instance, be sure to leave intermediate results on the remote nodes in the cluster case, and in the GPU memory in that setting, if possible.

2.6 Thread Scheduling

Say you have a threaded program, for example with four threads and a machine with four cores. Then the four threads will run physically simultaneously (if there are no other programs competing with them). That of course is the entire point, to achieve parallelism, but there is more to it than that.

Modern operating systems for general-purpose computers use *timesharing*: Unseen by the users, programs are taking turns (*timeslices*) using the cores of the machine. Say for instance that Manny and Moe are using a university computer named Jack, with Manny sitting at the console and Moe logged in remotely. For concreteness, say Manny is running an R program, and Moe is running something in Python, with both currently having long-running computations in progress.

Assume first that this is a single-core machine. Then only one program can run at a time. Manny's program will run for a while, but after a set amount of time, the hardware timer will issue an *interrupt*, causing a jump to another program. That program has been configured to be the operating system. The OS will look on its *process table* to find another program in *ready* state, meaning runnable (as opposed to say, suspended while awaiting keyboard input). Assuming there are no other processes, Moe's program will now get a turn. This transition from Manny to Moe is called a *context switch*. Moe's program will run for a while, then another interrupt comes, and Manny will get another turn, and so on.

Now suppose it is a dual-core machine. Here Manny and Moe's programs will run more or less continuously, in parallel, though with periodic downtimes due to the interrupts and attendant brief OS runs.

But suppose Moe's code is threaded, running two threads. Now we will have three threads—Moe's two and Manny's one (even a non-threaded program consists of one thread)—competing to use three cores. Moe's two threads will sometimes run in parallel with each other but sometimes not. Instead of a 2X speedup, Moe is getting about 1.5X.[10]

There are also possible cache issues. When a thread starts a new turn, it may be on a different core than that used in the last turn. If there is a separate cache for each core, the cache as the new core probably contains little if anything useful to this thread. Thus there will be a lot of cache misses for a while in this timeslice. There may be a remedy in the form of setting *processor affinity*; see Section 5.9.4.

By the way, what happens when one of those programs finishes its computation and returns to the user prompt, e.g., > in the case of Manny's R program? R will then be waiting for Manny's keyboard input. But the OS won't wait, and the OS does in fact get involved. R is trying to read from the keyboard, and to do this it calls a C library function, which in turn makes a call to a function in the OS. The OS, realizing that it may be quite a while before Manny types, will mark his entry in the process table as being in *sleep* state. When he finally does hit a key, the keyboard sends an interrupt,[11] causing the OS to run, and the latter will mark his program as now being back in ready state, and it will eventually get another turn.

[10]Even the 2X figure assumes that Moe's code was load balanced in the first place, which may not be the case.

[11]In Moe's case, the interrupt will come from Jack's network card.

2.7 How Many Processes/Threads?

As mentioned earlier, it is customary in the R world to refer to each worker in a **snow** program as a process. A question that then arises is, how many processes should we run?

Say for instance we have a cluster of 16 nodes. Should we set up 16 workers for our **snow** program? The same issues arise with threaded programs, say with **Rdsm** or OpenMP (Chapters 4 and 5). On a quadcore machine, should we run 4 threads?

The answer is *not* automatically Yes to these questions. With a fine-grained program, using too many processes/threads may actually degrade performance, as the overhead may overwhelm the presumed advantage of throwing more hardware at the problem. So, one might actually use fewer cluster nodes or fewer cores than one has available.

On the other hand, one might try to *oversubscribe* the resources. As discussed earlier, a cache miss causes a considerable delay, and a page fault even more. This is time during which one of the nodes/cores will not be doing any computation, exacting an opportunity cost from performance. It may pay, then, to have "extra" threads for the program available to run.

2.8 Example: Mutual Outlink Problem

To make this concrete, let's measure times for the mutual outlinks problem (Section 1.4), with larger and larger numbers of processes.

Here I ran on a shared-memory machine consisting of four processor chips, each of which has eight cores. This gives us a 32-core system, and I ran the mutual outlinks problem with values of **nc**, the number of cores, equal to 2, 4, 6, 8, 10, 12, 16, 24, 28 and 32. The problem size was 1000 rows by 1000 columns. The times are plotted in Figure 2.2.

Here we see a classical U-shaped pattern: As we throw more and more processes on the problem, it helps in the early stages, but performance actually degrades after a certain point. The latter phenomenon is probably due to the communications overhead we discussed earlier, in this case bus contention and the like.[12]

[12]Though the processes are independent and do not share memory, they do share the bus.

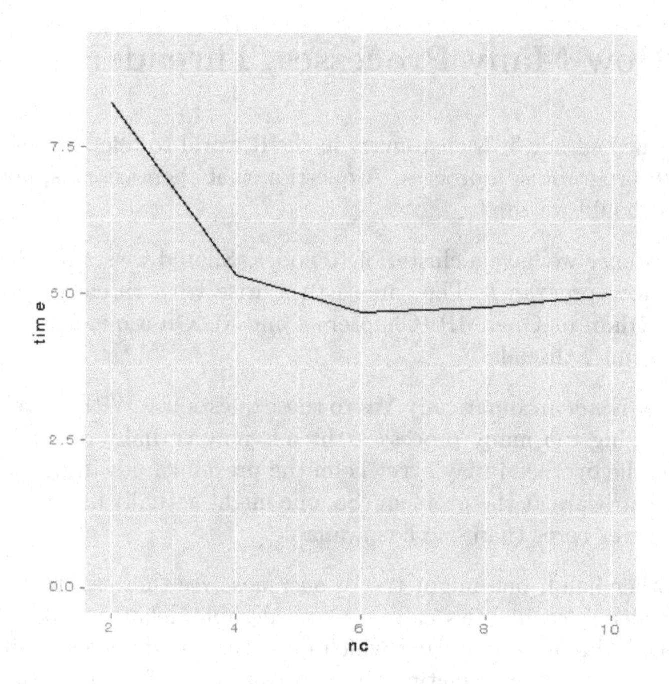

Figure 2.2: Run Time Versus Number of Cores

By the way, for each of our **nc** workers, we had one invocation of R running on the machine. There was also an additional invocation, for the manager. However, this is not a performance issue in this case, as the manager spends most of its time idle, waiting for the workers.

2.9 "Big O" Notation

With all this talk of physical obstacles to overcome, such as memory access time, it's important also to raise the question as to whether the application itself is very parallelizable in the first place. One measure of that is "big O" notation.

In our mutual outlinks example with an $n \times n$ adjacency matrix, we need to do on average $n/2$ sum operations per row, with n rows, thus $n \cdot n/2$ operations in all. In parallel processing circles, the key question asked about hardware, software, algorithms and so on is, "Does it scale?", meaning, Does the run time grow manageably as the problem size grows?

We see above that the run time of the mutual outlinks problem grows proportionally to the *square* of the problem size, in this case the number of websites. (Dividing by 2 doesn't affect this growth rate.) We write this as $O(n^2)$, known colloquially as "big O" notation. When applied to analysis of run time, we say that it measures the *time complexity*.

Ironically, applications that *are* manageable often are poor candidates for parallel processing, due to overhead playing a greater role in such problems. An application with $O(n)$ time complexity, for instance, may present a challenge. We will return to this notion at various points in this book.

2.10 Data Serialization

Some parallel R packages, e.g., **snow**, that send data through a network *serialize* the data, meaning to convert it to ASCII form. The data must then be unserialized on the receiving end. This creates a delay, which may or may not be serious but must be taken into consideration.

2.11 "Embarrassingly Parallel" Applications

The term *embarrassingly parallel* is heard often in talk about parallel programming. It is a central topic, hence deserving of having a separate section devoted to it.

2.11.1 What People Mean by "Embarrassingly Parallel"

It's no shame to be poor...but it's no great honor either—the character Tevye in *Fiddler on the Roof*

Consider a matrix multiplication application, for instance, in which we compute AX for a matrix A and a vector X. One way to parallelize this problem would be to have each processor handle a group of rows of A, multiplying each by X in parallel with the other processors, which are handling other groups of rows. We call the problem embarrassingly parallel, with the word "embarrassing" meaning that the problem is too easy, i.e., there is no intellectual challenge involved. It is pretty obvious that the computation Y = AX can be parallelized very easily by splitting the rows of A into groups.

By contrast, most parallel sorting algorithms require a great deal of interaction. For instance, consider Mergesort. It breaks the vector to be sorted into two (or more) independent parts, say the left half and right half, which are then sorted in parallel by two processes. So far, this is embarrassingly parallel, at least after the vector is broken in half. But then the two sorted halves must be merged to produce the sorted version of the original vector, and that process is *not* embarrassingly parallel; it can be parallelized, but in a more complex, less obvious manner.

Of course, it's no shame to have an embarrassingly parallel problem! On the contrary, except for showoff academics, having an embarrassingly parallel application is a cause for celebration, as it is easy to program.

In recent years, the term *embarrassingly parallel* has drifted to a somewhat different meaning. Algorithms that are embarrassingly parallel in the above sense of simplicity tend to have very low communication between processes, key to good performance. That latter trait is the center of attention nowadays, so the term *embarrassingly parallel* generally refers to an algorithm with low communication needs.

2.11.2 Suitable Platforms for Non-Embarrassingly Parallel Applications

The only general-purpose parallel computing platform suitable for non-embarrassingly parallel applications is that of the multicore/multiprocessor system. This is due to the fact that processor/memory copies have the least communication overhead. Note carefully that this does not mean there is NO overhead—if a cache coherency transaction occurs, we pay a heavy price. But at least the "base" overhead is small.

Still, non-embarrassingly parallel problems are generally tough nuts to crack. A good, commonplace example is linear regression analysis. Here a matrix inversion or equivalent such as QR factorization, is tough to parallelize. We'll return to this issue frequently in this book.

2.12 Further Reading

For more information on memory addresses and structures, timesharing operating systems and so on, see my open source book on computer systems, *An Introduction to Computer Systems*, N. Matloff, `http://heather.cs.ucdavis.edu/~matloff/50/PLN/CompSystsBook.pdf`.

Chapter 3

Principles of Parallel Loop Scheduling

Many applications of parallel programming, both in R and in general, involve the parallelization of **for** loops. As will be explained shortly, this at first would seem to be a very easily programmed class of applications, but there can be serious performance issues.

First, though, let's define the class under consideration. Throughout this chapter, it will be assumed that the iterations of a loop are independent of each other, meaning that the execution of one iteration does not use the results of a previous one.

Here is an example of code that does not satisfy this condition:

```
total <- 0
for (i in 1:n) total <- total + x[i]
```

Putting aside the fact that this computation can be done with R's **sum()** function, the point is that for each **i**, the computation needs the previous value of **total**.

With this restriction of independent iterations, it would seem that we have an embarrassingly parallel class of applications. In terms of programmability, it is true. Using **snow**, for example in the mutual Web links code in Section 1.4.5, we simply called **clusterApply()** on the range of **i** that we had had in our serial loop:

```
ichunks <- 1:(nr-1)
```

```
tots <- clusterApply(cls,ichunks,doichunk)
```

This distributed the various iterations for execution by the workers. So, isn't it equally simple for any **for** loop?

The answer is No, because different iterations may have widely different times. If we are not careful, we can end up with a serious load balance issue. In fact, this was even the case in the mutual Web links code above—for larger values of **i**, the function **doichunk()** has less work to do: In the (serial) code in Listing 1.4.1, page 6, the matrix multiplication involves a matrix with **n-i** rows at iteration **i**.

This can cause big load balancing problems if we are not careful as to how we assign iterations to workers, i.e., how we do the loop scheduling. Moreover, we typically don't know the loop iteration times in advance, so the problem of efficient loop scheduling is even more difficult. Methods to address these issues will be the thrust of this chapter.

3.1 General Notions of Loop Scheduling

Suppose we have k processes and many loop iterations. Suppose too that we do not know beforehand how much time each loop iteration will take. Common types of loop scheduling are the following:

- *Static* scheduling: The assignment of loop iterations to processes is arranged before execution starts.
- *Dynamic* scheduling: The assignment of loop iterations to processes is arranged during execution. Each time a process finishes a loop iteration, it picks up a new one (or several, with chunking) to work on.
- *Chunking:* Assign a group of loop iterations to a process, rather than a single loop iteration. In dynamic scheduling, say, when a process becomes idle, it picks up a group of loop iterations to work on next.
- *Reverse* scheduling: In some applications, the execution time for an iteration grows larger as the loop index grows. For reasons that will become clear below, it is more efficient to reverse the order of the iterations.

Note that while static and dynamic scheduling are mutually exclusive, one can do chunking and reverse scheduling with either.

To make this concrete, suppose we have loop iterations A, B and C, and have two processes, P_1 and P_2. Consider two loop schedules:

- **Schedule I:** Dole out the loop iterations in *Round Robin*, i.e., cyclic order—assign A to P_1, B to P_2 and C to P_1, statically..
- **Schedule II:** Dole out the loop iterations dynamically, one at a time, as execution progresses. Let us suppose we do this in reverse order, i.e., C, B and A, because we suspect that their loop iteration times decrease in this order. (The relevance of this will be seen below.)

Now suppose loop iterations A, B and C have execution times of 10, 20 and 40, respectively. Let's see how early we would finish the full loop iteration set, and how much wasted idleness we would have, under both schedules.

In Schedule I, when P_1 finishes loop iteration A at time 10, it starts C, finishing the latter at time 50. P_2 finishes at time 20, and then sits idle during time 20-50.

Under Schedule II, there may be some randomness in terms of which of P_1 and P_2 gets loop iteration C. Say it is P_1. P_1 will execute only loop iteration C, never having a chance to do more. P_2 will do B, then pick up A and perform that loop iteration. The overall loop iteration set will be completed at time 40, with only 10 units of idle time. In other words, Schedule II outperforms Schedule I, both in terms of how soon we complete the project and how much idle time we must tolerate.

By the way, note that a static version of Schedule II, still using the (C,B,A) order, would in this case have the same poor performance as Schedule I.

There are two aspects, though, which we must consider:

- As mentioned earlier, typically we do not know the loop iteration times in advance. In the above example, we had loop iterations in Schedule II get their work in reverse order, due to a suspicion that C would take the longest etc. That guess was correct (in this contrived example), and placing our work queue in reverse order like that turned out to be key to the superiority of Schedule II in this case.
- Schedule II, and any dynamic method, may exact a substantial overhead penalty. In **snow**, for instance, there would need to be communication between a worker and the manager, in order for the worker to determine which task is next assigned to it. Static scheduling doesn't have this drawback.

This is the motivation for chunking in the dynamic case (though it can be used in the static case too). By assigning loop iterations to processes in groups instead of singly, processes need to go to the work queue less often, thus accruing less overhead.

On the other hand, large chunk sizes potentially bring back the problem of load imbalance. The final chunk handled by each process may begin at substantially different times from one process to another. This results in some processes incurring idle time—exactly the problem dynamic scheduling was meant to ameliorate. Thus some scheduling methods have been developed in which the chunk size decreases over time, saving overhead early in the computation, but reducing the possibility of substantial load imbalance near the end. (More on this in Section 5.3.)

3.2 Chunking in snow

The **snow** package itself doesn't provide a chunking capability. This is easily handled on one's own, though, which will be seen in our revised version of our mutual outlinks code.

3.2.1 Example: Mutual Outlinks Problem

Only one line of the code from Section 1.4.5 will be changed, but for convenience let's see it all in one piece:

```
doichunk <- function(ichunk) {
   tot <- 0
   nr <- nrow(lnks)  # lnks global at worker
   for (i in ichunk) {
      tmp <- lnks[(i+1):nr,] %*% lnks[i,]
      tot <- tot + sum(tmp)
   }
   tot
}

mutoutpar <- function(cls,lnks) {
   nr <- nrow(lnks)
   clusterExport(cls,"lnks")
   ichunks <- clusterSplit(cls,1:(nr-1))
   tots <- clusterApply(cls,ichunks,doichunk)
```

```
    Reduce(sum, tots) / nr
}
```

As before, our function **mutoutpar()** divides the **i** values into chunks, but now they are real chunks, not one **i** value per chunk as before. It does so via the **snow** function **clusterSplit()**:

```
mutoutpar <- function(cls, lnks) {
    nr <- nrow(lnks)  # lnks global at manager
    clusterExport(cls,"lnks")
    ichunks <- clusterSplit(cls,1:(nr-1))
    tots <- clusterApply(cls,ichunks,doichunk)
    Reduce(sum, tots) / nr
}
```

So, what does **clusterSplit()** do? Say **lnks** has 500 rows and we have 4 workers. The goal here is to partition the row numbers 1,2,...,500 into 4 equal (or roughly equal) subsets, which will serve as the chunks of indices for each worker to process. Clearly, the result should be 1-125, 126-250, 251-375 and 376-500, which will correspond to the values of **i** in the outer **for** loop in our serial code, Listing 1.4.1. Worker 1 will process the outer loop iterations for **i** = 1,2,...,125, and so on.

Let's check this. To save space below, let's try it on a smaller example, 1,2,...,50, on the cluster **cls**:

```
> clusterSplit(cls,1:50)
[[1]]
 [1]  1  2  3  4  5  6  7  8  9 10 11 12 13

[[2]]
 [1] 14 15 16 17 18 19 20 21 22 23 24 25

[[3]]
 [1] 26 27 28 29 30 31 32 33 34 35 36 37

[[4]]
 [1] 38 39 40 41 42 43 44 45 46 47 48 49 50
```

The call to **clusterSplit()** returned a list with 4 elements, each of which is a vector showing the indices to be processed by a given worker. It did work as expected. Since 50 is not divisible by 4, **snow** gave me subsets of sizes 13, 12, 12 and 13. The function tries to make the subsets as evenly divided as possible.

So, again thinking of the case of 500 rows and 4 workers, the code

```
ichunks <- clusterSplit(cls,1:(nr-1))
tots <- clusterApply(cls,ichunks,doichunk)
```

will send the chunk 1:125 to the first worker, 126:250 to the second, 251:375 to the third, and 375:499 to the fourth. The return list, assigned to **tots**, will now consist of four elements, rather than 499 as before.

Again, the only change from the previous version of this code was to add real chunks. This ought to help, because it allows us to better leverage the fact that R can do matrix multiplication fairly quickly. Let's see if this turns out to be the case. Here is a brief timing experiment using 8 cores on our usual 32-core machine, on a 1000 × 1000 problem:

chunking?	time
no	9.062
yes	6.264

Indeed, we got a speed improvement of about 30%.

3.3 A Note on Code Complexity

In general, chunking reduces overhead. This does also mean an increase in code complexity in many cases, but it can be very much worthwhile. For instance, in the example in Section 3.4.5, we find that a nonchunked version runs more slowly than the serial code, while the chunked version has much greater speed than the serial one.

Thus the code from here on will sometimes be more complex than what we have seen before. The algorithms themselves are usually simple, but the implementation often involves a lot of detail.

Welcome to the world of parallel programming! Working with details is a fact of life for such programming. But as long as you keep your eye on the big picture—the main points in the strategy in the design of the code—you'll have no trouble following the examples here, and more importantly, writing your own code. You need not be a professional programmer to write good parallel code; you simply need patience.

On a related note, the reader may be aware of the fact that **for** loops are generally avoided by experienced R programmers. In some cases this is to achieve better speed, but in others the goal is simply to write compact code, which tends to be easier to read (though often less easy to write). But the reader should not hesitate to make liberal use of **for** loops when the main advantage of non-loop code would be code compactness. In particular, use of **apply()** typically does not bring a speed improvement, and though we use it frequently in this book, the reader may prefer to stick with good old-fashioned loops instead.

3.4 Example: All Possible Regressions

Consider linear regression analysis, one of the mainstays of statistical methodology. Here one tries to predict one variable from others.

A major issue is the choice of predictor variables: On the one hand, one wants to include all relevant predictors in the regression equation. But on the other hand, we must avoid overfitting, and complex models may be hard to interpret. Thus a nice, compact, *parsimonious* equation is desirable.

Suppose we have n observations and p predictor variables. In the *all possible regressions* method of variable selection, we fit regression models to each possible subset of the p predictors, and choose the one we like best according to some criterion. The one we'll use in our example here is *adjusted* R^2, a (nearly) statistically unbiased estimator of the (population value of the) traditional R^2 criterion. In other words, we will choose for our model the predictor set for which adjusted R^2 is largest.

3.4.1 Parallelization Strategies

There are 2^p possible models, so the amount of computation could be quite large—a perfect place to use parallel computation. There are two possibilities here:

(a) For each set of predictors, we could perform the regression computation for that set in parallel. For instance, all the processes would work in concert in computing the model using predictors 2 and 5.

(b) We could assign a different collection of predictor sets to each process, with the process then performing the regression computations for those assigned sets. So, for example, one process might do the entire computation for the model involving predictors 2 and 5, another process would handle the model using predictors 8, 9 and 12, and so on.

Option (a) has problems. For a given set of m predictors, we must first compute various sums of squares and products. Each sum has n summands, and there are $O(m^2)$ sums, making for a computational complexity of $O(nm^2)$. (Recall that this notation was introduced in Section 2.9.) Then a matrix inversion (or equivalent operation, such as QR factorization) must be done, with complexity $O(m^3)$.[1]

Unfortunately, matrix inversion is not an embarrassingly parallel operation, and though many good methods have been developed, it is much easier here to go the route of option (b). The latter *is* embarrassingly parallel, and in fact involves a loop.

Below is a **snow** implementation of doing this in parallel. It finds the adjusted R^2 value for all models in which the predictor set has size at most k. The user can opt for either static or dynamic scheduling, or reverse the order of iterations, and can specify a (constant) chunk size.

3.4.2 The Code

```
# regresses response variable Y column against
# all possible subsets of the Xi predictor variables,
# with subset size up through k; returns the
# adjusted R-squared value for each subset

# scheduling parameters:
#
#   static (clusterApply())
#   dynamic (clusterApplyLB())
#   reverse the order of the tasks
#   chunk size (in dynamic case)

# arguments:
#     cls: Snow cluster
#     x: matrix of predictors, one per column
#     y: vector of the response variable
#     k: max size of predictor set
#     reverse: TRUE means reverse the order
#                of the iterations
#     dyn: TRUE means dynamic scheduling
#     chunksize: scheduling chunk size
```

[1] In the QR case, the complexity may be $O(m^2)$, depending on exactly what is being computed.

```
# return value:
#    R matrix, showing adjusted R-squared values,
#        indexed by predictor set

snowapr <- function(cls,x,y,k,reverse=FALSE,dyn=FALSE,
    chunksize=1) {
  require(parallel)
  p <- ncol(x)
  # generate predictor subsets, an R list,
  # 1 element for each predictor subset
  allcombs <- genallcombs(p,k)
  ncombs <- length(allcombs)
  clusterExport(cls,"do1pset")
  # set up task indices
  tasks <- if (!reverse)
      seq(1,ncombs,chunksize) else
    seq(ncombs,1,-chunksize)
  if (!dyn) {
    out <- clusterApply(cls,tasks,dochunk,x,y,
      allcombs,chunksize)
  } else {
    out <- clusterApplyLB(cls,tasks,dochunk,x,y,
      allcombs,chunksize)
  }
  # each element of out consists of rows showing
  # adj. R2 and the indices of the predictor set
  # that produced it; combine all those vectors
  # into a matrix
  Reduce(rbind,out)
}

# generate all nonempty subsets of 1..p of size <= k;
# returns an R list, one element per predictor set,
# in the form of a vector of indices
genallcombs <- function(p,k) {
  allcombs <- list()
  for (i in 1:k) {
    tmp <- combn(1:p,i)
    allcombs <- c(allcombs,matrixtolist(tmp,rc=2))
  }
  allcombs
}
```

```
# extracts rows (rc=1) or columns (rc=2) of a matrix,
# producing a list
matrixtolist <- function(rc,m) {
   if (rc == 1) {
      Map(function(rownum) m[rownum,],1:nrow(m))
   } else Map(function(colnum) m[,colnum],1:ncol(m))
}

# process all the predictor sets in the allcombs
# chunk whose first index is psetsstart
dochunk <- function(psetsstart,x,y,allcombs,
      chunksize) {
   ncombs <- length(allcombs)
   lasttask <- min(psetsstart+chunksize-1,ncombs)
   t(sapply(allcombs[psetsstart:lasttask],
      do1pset,x,y))
}

# find the adjusted R-squared values for the given
# predictor set onepset; return value will be the
# adj. R2 value, followed by the predictor set
# indices, with 0s as filler--for convenience, all
# vectors returned by calls to do1pset() have
# length k+1; e.g. for k = 4, (0.28,1,3,0,0) would
# mean the predictor set consisting of columns 1 and
# 3 of x, with an R2 value of 0.28
do1pset <- function(onepset,x,y) {
   slm <- summary(lm(y ~ x[,onepset]))
   n0s <- ncol(x) - length(onepset)
   c(slm$adj.r.squared,onepset,rep(0,n0s))
}

# predictor set seems best
snowtest <- function(cls,n,p,k,chunksize=1,
      dyn=FALSE,rvrs=FALSE) {
   gendata(n,p)
   snowapr(cls,x,y,k,rvrs,dyn,chunksize)
}

gendata <- function(n,p) {
   x <<- matrix(rnorm(n*p),ncol=p)
   y <<- x%*%c(rep(0.5,p)) + rnorm(n)
}
```

3.4.3 Sample Run

Here is some sample output:

```
> snowtest(c8,100,4,2)
            [,1] [,2] [,3] [,4] [,5]
 [1,] 0.21941625    1    0    0    0
 [2,] 0.05960716    2    0    0    0
 [3,] 0.11090411    3    0    0    0
 [4,] 0.15092073    4    0    0    0
 [5,] 0.26576805    1    2    0    0
 [6,] 0.35730378    1    3    0    0
 [7,] 0.32840075    1    4    0    0
 [8,] 0.17534962    2    3    0    0
 [9,] 0.20841665    2    4    0    0
[10,] 0.27900555    3    4    0    0
```

Here simulated data of size $n = 100$ was generated, with $p = 4$ predictors and a maximum predictor set size of $k = 2$. The highest adjusted R^2 value was about 0.36, for the model using predictors 1 and 3, i.e., columns 1 and 3 of **x**.

3.4.4 Code Analysis

As noted in Section 3.3, parallel code does tend to involve a lot of detail, so it is important to keep in mind the overall strategy of the algorithm. In the case at hand here, the strategy is as follows:

- The manager determines all the predictor sets of size up to k.
- The manager assigns each worker to handle specified predictor sets.
- Each worker calculates the adjusted R^2 value for each of its assigned predictor sets.
- The manager collects the results, and assembles them into a results matrix. The ith row of the matrix shows the adjusted R^2 values and their associated predictor sets.

Note that our approach here is consistent with the discussion in Section 1.1, i.e., to have our code leverage the power of R: Each worker calls the R linear model function **lm()**.

To understand the details, in the following continue to consider the case of $p = 4$, $k = 2$. Also, suppose our chunk size is 2, and we have two workers. We will use static, nonreverse scheduling.

3.4.4.1 Our Task List

Our main function **snowapr()** will first call **genallcombs()** which, as its name implies, will generate all the combinations of predictor variables, one combination per list element:

```
> genallcombs(4,2)
[[1]]
[1] 1

[[2]]
[1] 2

[[3]]
[1] 3

[[4]]
[1] 4

[[5]]
[1] 1 2

[[6]]
[1] 1 3

[[7]]
[1] 1 4

[[8]]
[1] 2 3

[[9]]
[1] 2 4

[[10]]
[1] 3 4
```

For example, the last list element says that one of the combinations is (3,4), corresponding to the model with predictors 3 and 4, i.e., columns 3 and 4 of **x**.

Thus, the list **allcombs** is our task list, one task per element of the list.

As mentioned, the basic idea is simple: We distribute these tasks, 10 of them in this case, to the workers. Each worker then runs regressions on each of its assigned combinations, and returns the results to the manager, which coalesces them.

3.4.4.2 Chunking

Here we set up chunking, with the line

```
tasks <- seq(1,ncombs,chunksize)
```

In the above example, **tasks** will be (1,3,5,7,9). Our code will interpret these numbers as the starting indices of the various chunks, with for example 3 meaning the chunk starting at the third combination, i.e., the third element of **allcombs**. Since our chunk size is 2 in this example, the chunk will consist of the third and fourth combinations in **allcombs:** This chunk will consist of two single-predictor models, one using predictor 3 and the other using predictor 4.

3.4.4.3 Task Scheduling

Let us name our two workers P_1 and P_2, and suppose we use static scheduling, the default for **snow**. The package implements scheduling in a *Round Robin* manner. Recalling that our vector **tasks** is (1,3,5,7,9), we see that 1 will be assigned to P_1, 3 will be assigned to P_2, 5 will be assigned to P_1, and so on. Again, note that assigning 3 to P_2, for instance, means that combinations 3 and 4 will be handled by that worker, since our chunk size is 2.

In our call to **snowapr()**, we would set **chunksize** to 2 and set **dyn** to FALSE, as we are using static scheduling. We are not reversing the order of tasks, so we set **rvrs** to FALSE.

In the dynamic case, at first the assignment will match the static case, with P_1 getting combinations 1 and 2, and P_2 being assigned 3 and 4. After that, though, things are unpredictable. The manager could assign combinations 5 and 6 to either P_1 or P_2, depending on which worker finishes its initial combinations first. It's a "first come, first served" kind of setup. The **snow** package includes a variant of **clusterApply()** that does dynamic scheduling, named **clusterApplyLB()** ("LB" for "load balance").

As seen in the toy example in Section 3.1, it may be advantageous to schedule iterations in reverse order. This is requested by setting **reverse** to TRUE. Since iteration times are clearly increasing in this application, we should consider using this option.

3.4.4.4 The Actual Dispatching of Work

That brings us to the heart of the code, the **snow** call

```
out <- clusterApply(cls,tasks,dochunk,x,y,allcombs,
      chunksize)
```

(and the paired call to **clusterApplyLB()**, which works the same way). As mentioned, **tasks** will be (1,3,5,7,9), each element of which will be fed into the function **dochunk()** by a worker. P_1, as noted, will do this for the elements 1, 5 and 9, resulting in three calls to **dochunk()** being made by P_1. In those calls, **psetsstart** will be set to 1, 5 and 9, respectively.

Note that we've written our function **dochunk()** to have five arguments. The first one will come from a portion of **tasks**, as explained above. The value of that argument will be different for each worker. But the other four arguments will be taken from the items that follow **dochunk** in the call

```
out <- clusterApply(cls,tasks,dochunk,x,y,allcombs,
          chunksize)
```

The values of these arguments will be the same for all workers. The **snow** function **clusterApply()** is structured this way, i.e., with all arguments following the worker function (**dochunk()** in this case) being assigned in common by all workers.

For convenience, here is a copy of the code of relevance right now:

```
dochunk <- function(psetsstart,x,y,allcombs,
      chunksize) {
   ncombs <- nrow(allcombs)
   lasttask <- min(psetsstart+chunksize-1,ncombs)
   t(sapply(allcombs[psetsstart:lasttask],
         do1pset,x,y))
}
```

```
do1pset <- function(onepset,x,y){
   slm <- summary(lm(y ~ x[,onepset]))
   n0s <- ncol(x) - length(onepset)
   c(slm$adj.r.squared,onepset,rep(0,n0s))
}
```

And here again is (part of) what we found earlier for **allcombs**:

```
[[1]]
[1] 1

[[2]]
[1] 2

[[3]]
[1] 3
...
```

Let's look at what happens when P_1 calls **dochunk()** on the 1 element, i.e., with **psetsstart** set to 1:

The name **psetsstart** is meant to evoke "predictor sets start," alluding to the fact that our predictor sets here start at element 1 of **allcombs**, in which the predictor set is just the singleton predictor 1, since **allcombs[[1]]** is just (1). And since **lasttask**, computed in the call to **min()**, will be 2, our second and last predictor set will be the singleton 2. To recap: P_1's work on the current chunk will consist of first performing a regression analysis using column 1 of **x** as a predictor, and then running a regression using column 2 instead.

Now let's look at the call to **sapply()** in **dochunk()**,

```
t(sapply(allcombs[psetsstart:lasttask],do1pset,x,y))
```

The specifies that **do1pset()** will first be called on **allcombs[psetsstart]**, then on **allcombs[psetsstart+1]** etc., up through **allcombs[lasttask]**. In other words, **do1pset()** will be called on each predictor set in this worker's chunk of **allcombs**. In the case at hand, this will be the set {1} and the set {2}.

Since the return value from **do1pset()** has a vector type, the results of **sapply()** will be arranged in columns. Thus in the end a call to the matrix transpose function **t()** is needed.

The function **do1pset()** itself is fairly straightforward. Note that one of the components of the object returned by the call to the regression function **lm()** and then **summary()** is **adj.r.squared**, the adjusted R^2 value.

The end result will be that the call to **dochunk()** with **psetsstart** equal to 1 will return rows 1 and 2 of the final output seen in Section 3.4.3. Thus chunking is handled in this manner, in spite of the lack of a chunking capability in **snow** itself.

That's quite a bit to digest! The partitioning of work due to chunking was rather intricate, and a nonchunked version would have been much simpler. But we will find in Section 3.4.5, the chunking is necessary; without it, our parallel code would be slower than the serial version.

3.4.4.5 Wrapping Up

Back in **snowapr()**, we use **Reduce()** to amalgamate the results returned by the workers (which, as before, will be in list form):

```
Reduce(rbind,out)
```

3.4.5 Timing Experiments

No attempt will be made here to do an exhaustive analysis, varying all the factors—n, p, the scheduling methods, chunk size, number of processes and so on. But let's explore a little.

Here are some timings with $n = 10000$, $p = 20$ and $k = 3$ on our usual 32-core machine, though only eight cores were used here. As a baseline, let's see how long a run takes with just one core (without using **snow**). A modified version of the code (not shown), yields the following:

```
> system.time(apr(x,y,3))
   user  system elapsed
 35.070   0.132  35.295
```

Now let's try it on a two-process cluster:

```
> system.time(snowapr(c2,x,y,3))
   user  system elapsed
 31.006   5.028  77.447
```

This is awful! Instead of cutting the run time in half, using two processes actually doubled the time. This is a great example of the problems that overhead can bring.

Let's see if dynamic scheduling helps:

```
> system.time(snowapr(c2,x,y,3,dyn=TRUE))
   user  system elapsed
 33.370   4.844  64.543
```

A little better, but still slower than the serial version. Maybe chunking will help?

```
> system.time(snowapr(c2,x,y,3,dyn=TRUE,chunk=10))
   user  system elapsed
  2.904   0.572  22.753
> system.time(snowapr(c2,x,y,3,dyn=TRUE,chunk=25))
   user  system elapsed
  1.340   0.240  19.677
> system.time(snowapr(c2,x,y,3,dyn=TRUE,chunk=50))
   user  system elapsed
  0.652   0.128  19.692
```

Ah! That's more like it. It's not quite clear from this limited experiment what chunk size is best, but all of the above sizes worked well.

How about an eight-process **snow** cluster?

```
> system.time(snowapr(c8,x,y,3,dyn=TRUE,chunk=10))
   user  system elapsed
  3.861   0.568   7.542
> system.time(snowapr(c8,x,y,3,dyn=TRUE,chunk=15))
   user  system elapsed
  2.592   0.284   6.828
> system.time(snowapr(c8,x,y,3,dyn=TRUE,chunk=20))
   user  system elapsed
  1.808   0.316   6.740
> system.time(snowapr(c8,x,y,3,dyn=TRUE,chunk=25))
   user  system elapsed
  1.452   0.232   7.082
```

This is approximately a five-fold speedup over the serial version, very nice.

Of course, theoretically we might hope for an eight-fold speedup, since we have eight processes, but overhead prevents that.

By the way, in thinking about the chunk size, it might be useful to check how many predictor sets we need to do in all:

```
> length(genallcombs(20,3))
[1] 1350
```

3.5 The partools Package

Recall the function **matrixtolist()** in the last section, which converts a matrix to an R list of the rows or columns of the matrix. Clearly this function would be useful in many other contexts. Thus I've collected this and various other functions/code snippets in a CRAN package, **partools**.

3.6 Example: All Possible Regressions, Improved Version

We did get good speedups above from parallelization, but at the same time we should have some nagging doubts. After all, we are doing an awful lot of duplicate work.

If you have background in the mathematics of linear models (don't worry about this if you don't, as the following will still be readable)), you know that the vector of estimated regression coefficients is calculated as

$$\widehat{\beta} = (X'X)^{-1}X'Y \tag{3.1}$$

(again, or with something like a QR decomposition instead of matrix inversion) where X is the matrix of predictor data (one column per predictor), Y is the vector of response variable values, and the prime symbol means matrix transpose. If we include a constant term in the model, as is standard, the first column of X consists of all 1s.

The problem is that in each of the calls to **lm()**, we are redoing part of this computation. In particular, look at the quantity $X'X$. For each set of predictors we use, we are forming this product for a different set of columns of X. Why not just do it once for all of X?

For example, say we are currently working with the predictor set (2,3,4). Let $\tilde{X}$ denote the analog of X for this set. Then it can be shown that $\tilde{X}'\tilde{X}$ is equal to the 3x3 submatrix of $X'X$ corresponding to rows 3-5 and columns 3-5 of the latter.

So it makes sense to calculate $X'X$ once and for all, and then extract submatrices as needed.

3.6.1 Code

```
# regresses response variable Y column against
# all possible subsets of the Xi predictor variables,
# with subset size up through k; returns the
# adjusted R-squared value for each subset

# this version computes X'X and X'Y first, and
# stores it at the workers

# scheduling methods:
#
#    static (clusterApply())
#    dynamic (clusterApplyLB())
#    reverse the order of the tasks
#    varying chunk size (in dynamic case)

# arguments:
#     cls:   cluster
#     x:   matrix of predictors, one per column
#     y:   vector of the response variable
#     k:   max size of predictor set
#     reverse:  TRUE means reverse the order of
#               the iterations
#     dyn:  TRUE means dynamic scheduling
#     chunksize:  scheduling chunk size
# return value:
#     R matrix, showing adjusted R-squared values,
#     indexed by predictor set

snowapr1 <- function(cls,x,y,k,reverse=FALSE,dyn=FALSE,
      chunksize=1) {
   require(parallel)
   # add 1s column
```

```
   x <- cbind(1,x)
   xpx <- crossprod(x,x)
   xpy <- crossprod(x,y)
   p <- ncol(x) - 1
   # generate matrix of predictor subsets
   allcombs <- genallcombs(p,k)
   ncombs <- length(allcombs)
   clusterExport(cls,"do1pset1")
   clusterExport(cls,"linregadjr2")
   # set up task indices
   tasks <- if (!reverse)
      seq(1,ncombs,chunksize) else
      seq(ncombs,1,-chunksize)
   if (!dyn) {
      out <- mclapply(tasks,dochunk2,
         x,y,xpx,xpy,allcombs,chunksize)
   } else {
      out <- clusterApplyLB(cls,tasks,dochunk2,
         x,y,xpx,xpy,allcombs,chunksize)
   }
   Reduce(rbind,out)
}

# generate all nonempty subsets of 1..p of size <= k;
# returns a list, one element per predictor set
genallcombs <- function(p,k) {
   allcombs <- list()
   for (i in 1:k) {
      tmp <- combn(1:p,i)
      allcombs <- c(allcombs,matrixtolist(tmp,rc=2))
   }
   allcombs
}

# extracts rows (rc=1) or columns (rc=2) of a matrix,
# producing a list
matrixtolist <- function(rc,m) {
   if (rc == 1) {
      Map(function(rownum) m[rownum,],1:nrow(m))
   } else Map(function(colnum) m[,colnum],1:ncol(m))
}

# process all the predictor sets in the chunk
```

```
# whose first index is psetstart
dochunk2 <- function(psetstart,x,y,xpx,xpy,
      allcombs,chunksize) {
   ncombs <- length(allcombs)
   lasttask <- min(psetstart+chunksize-1,ncombs)
   t(sapply(allcombs[psetstart:lasttask],do1pset1,
      x,y,xpx,xpy))
}

# find the adjusted R-squared values for the given
# predictor set index
do1pset1 <- function(onepset,x,y,xpx,xpy) {
   ps <- c(1,onepset+1)  # account for constant term
   x1 <- x[,ps]
   xpx1 <- xpx[ps,ps]
   xpy1 <- xpy[ps]
   ar2 <- linregadjr2(x1,y,xpx1,xpy1)
   n0s <- ncol(x) - length(ps)
   # form the report for this predictor set; need
   # trailing 0s so as to form matrices of uniform
   # numbers of rows, to use rbind() in snowapr()
   c(ar2,onepset,rep(0,n0s))
}

# finds regression estimates "from scratch"
linregadjr2 <- function(x,y,xpx,xpy) {
   bhat <- solve(xpx,xpy)
   resids <- y - x %*% bhat
   r2 <- 1 - sum(resids^2)/sum((y-mean(y))^2)
   n <-nrow(x); p <- ncol(x) - 1
   1 - (1-r2) * (n-1) / (n-p-1)  # adj R2
}

# which predictor set seems best
snowtest1 <- function(cls,n,p,k,chunksize=1,
      dyn=FALSE,rvrs=FALSE) {
   gendata(n,p)
   snowapr(cls,x,y,k,rvrs,dyn,chunksize)
}

gendata <- function(n,p) {
   x <<- matrix(rnorm(n*p),ncol=p)
   y <<- x%*%c(rep(0.5,p)) + rnorm(n) }
```

3.6.2 Code Analysis

There are only a few changes from the previous code:

- As mentioned, typically regression models include a constant term, i.e., the β_0 in the model

$$\text{mean response} = \beta_0 + \beta_1 \text{ predictor1} + \beta_2 \text{ predictor2} + ... \quad (3.2)$$

  To accommodate this, the math underpinnings of regression require that a column of 1s be prepended to the X matrix. This is done via the line

```
x <- cbind(1,x)
```

  in **snowapr1()**.[2]

- Our predictor set indices, e.g. (2,3,4) above, must then be shifted accordingly in **do1pset()**, now named **do1pset1()** in this new code.

```
ps <- c(1,onepset+1)  # account for constant term
```

- Note that R's **crossprod()** function is used. Called on matrices A and B, it computes $A'B$.

- The function **linregadjr2()** computes adjusted R^2 from the mathematical definition. (The R function **lm.fit()** could not be used here, as it would not take advantage of our having already computed $X'X$ and $X'Y$.)

3.6.3 Timings

Let's run **snowapr1()** in the same settings we did earlier for **snowapr()**. Again, this is for $n = 10000$, $p = 20$ and $k = 3$, all with **dyn = TRUE**,

[2]The reader should note that the code here has not been optimized for good numerical properties, e.g. matrix condition number.

reverse = FALSE on an eight-node **snow** cluster.

chunksize	snowapr()	snowapr1()
1	39.81	63.67
10	7.54	6.16
15	6.83	4.60
20	6.74	3.39
25	7.08	3.13

Aside from an odd increase in the nonchunked case, there was a marked improvement. But there's more: Since the times still seemed to be decreasing at **chunksize = 25**, I tried some larger sizes:

chunksize	snowapr1()
50	1.632
75	1.026
150	0.726
200	0.633
350	0.683
500	0.831

So not only did it help to precompute $X'X$ and $X'Y$ in terms of improving corresponding earlier times, it also enables much better exploitation of chunking.

The reader might wonder whether it would pay to parallelize those computations, i.e., of $X'X$ and $X'Y$. The answer is no for the problem sizes seen above; the time for serial computation of those two matrices is already quite small, so overhead would produce a net loss of speed. However, it may be worthwhile on much larger problems.

3.7 Introducing Another Tool: multicore

As explained in Section 1.4.2, the **parallel** package was formed from two contributed R packages, **snow** and **multicore**. Now that we've seen how the former works, let's take a look at the latter. (Note that just as we have been using **snow** as a shorthand for "the portion of the **parallel** package that was adapted from **snow**," we'll do the same for **multicore**.)

As the name implies, **multicore** must be run on a multicore machine. Also, it's restricted to Unix-family operating systems, notably Linux and the

Macintosh's OS X. But with such a platform, you may find that **multicore** outperforms **snow**.[3]

3.7.1 Source of the Performance Advantage

Unix-family OSs include a *system call*, i.e., a function in the OS that application programmers can call as a service, named **fork()**. This is *fork* as in "fork in the road," rather than in "knife and fork." The image the term is meant to evoke is that of a process splitting into two.

What **multicore** does is call the OS **fork()**. The result is that if you call one of the **multicore** functions in the **parallel** package, you will now have two or more instances of R running on your machine! Say you have a quad core machine, and you set **mc.cores** to 4 in your call to the **multicore** function **mclapply()**. You will now have five instances of R running—your original plus four copies. (You can check this by running your OS' **ps** command.)

This in principle should fully utilize your machine in the current computation—four child R processes running on four cores. (The parent R process is dormant, waiting for the four children to finish.)

An absolutely key point is that initially the four child R processes will be exact copies of the parent. They will have the same values of your variables, as of the time of the forks. Just as importantly, initially the four children are actually *sharing* the data, i.e., are accessing the same physical locations in memory. (Note the word *initially* above; any changes made to the variables by a worker process will NOT be reflected at the manager or at the other workers.)

To see why that is so important, think again of the all possible regressions example earlier in this chapter, specifically the improved version discussed in Section 3.6. The idea there was to limit duplicate computation, by determining **xpx** and **xpy** just once, and sending them to the workers.

But the latter is a possible problem. It may take quite some time to send large objects to the workers. In fact, shipping the two matrices to the workers adds even more overhead, since as noted in Section 2.10, the **snow** package serializes communication.

But with **multicore**, no such action is necessary. Because **fork()** creates exact, *shared*, copies of the original R process, they all already have the variables **xpx** and **xpy**! At least for Linux, a *copy-on-write* policy is used,

[3] One should add, "In the form of **snow** used so far." More on this below.

which is to have the child processes physically share the data until such time as it is written to. But in this application, the variables do not change, so using **multicore** should be a win. Note that the same gain might be made for the variable **allcombs** too.

The **snow** package also has an option based on **fork()**, called **makeForkCluster()**. Thus, potentially this same performance advantage can be attained in **snow**, using that function instead of **makeCluster()**. If you are using **snow** on a multicore platform, you should consider this option.

3.7.2 Example: All Possible Regressions, Using multicore

The workhorse of **multicore** is **mclapply()**, a parallel version of **lapply()**. Let's convert our previous code to use this function. Since it is largely similar to **snow**'s **clusterApply()**, the changes to our previous code will be pretty minimal. In fact, since there are no (explicit) clusters, our code here will be somewhat simpler than the **snow** version.

Here's the code:

```
# regresses response variable Y column against
# all possible subsets of the Xi predictor variables,
# with subset size up through k; returns the
# adjusted R-squared value for each subset

# this version computes X'X and X'Y first

# scheduling methods:
#
#   static (clusterApply())
#   dynamic (clusterApplyLB())
#   reverse the order of the tasks
#   chunk size (in dynamic case)

# arguments:
#    x:  matrix of predictors, one per column
#    y:  vector of the response variable
#    k:  max size of predictor set
#    reverse:  TRUE means reverse the order
#              of the iterations
#    dyn:  TRUE means dynamic scheduling
#    chunk:  chunk size
```

```
# return value:
#    R matrix, showing adjusted R-squared values,
#    indexed by predictor set

mcapr <- function(x,y,k,ncores,reverse=FALSE,
      dyn=FALSE,chunk=1) {
   require(parallel)
   # add 1s column to X
   x <- cbind(1,x)
   # find X'X, X'Y
   xpx <- crossprod(x,x)
   xpy <- crossprod(x,y)
   # generate matrix of predictor subsets
   allcombs <- genallcombs(ncol(x)-1,k)
   ncombs <- length(allcombs)
   # set up task indices
   tasks <- if (!reverse) seq(1,ncombs,chunk) else
      seq(ncombs,1,-chunk)
   out <- mclapply(tasks,dochunk2,x,y,xpx,xpy,
      allcombs,chunk, mc.cores=ncores,mc.
         preschedule=!dyn)
   Reduce(rbind,out)
}

# process all the predictor sets in the chunk
# whose first allcombs index is psetsstart
dochunk2 <- function(psetsstart,x,y,
      xpx,xpy,allcombs,chunk) {
   ncombs <- length(allcombs)
   lasttask <- min(psetsstart+chunk-1,ncombs)
   t(sapply(allcombs[psetsstart:lasttask],do1pset2,
      x,y,xpx,xpy))
}

# find the adjusted R-squared values for the given
# predictor set, onepset
do1pset2 <- function(onepset,x,y,xpx,xpy) {
   ps <- c(1,onepset+1)  # account for 1s column
   xps <- x[,ps]
   xpxps <- xpx[ps,ps]
   xpyps <- xpy[ps]
   ar2 <- linregadjr2(xps,y,xpxps,xpyps)
   n0s <- ncol(x) - length(ps)
```

```
   # form the report for this predictor set; need
   # trailing 0s so as to form matrices of uniform
   # numbers of rows, to use rbind() in mcapr()
   c(ar2,onepset,rep(0,n0s))
}

# do linear regression with given xpx, xpy,
# return adj. R2
linregadjr2 <- function(xps,y,xpx,xpy) {
   # get beta coefficient estimates
   bhat <- solve(xpx,xpy)
   # find R2 and then adjusted R2
   resids <- y - xps %*% bhat
   r2 <- 1 - sum(resids^2)/sum((y-mean(y))^2)
   n <-nrow(xps); p <- ncol(xps) - 1
   1 - (1-r2) * (n-1) / (n-p-1)
}

# generate all nonempty subsets of 1..p of size <= k;
# returns a list, one element per predictor set
genallcombs <- function(p,k) {
   allcombs <- list()
   for (i in 1:k) {
      tmp <- combn(1:p,i)
      allcombs <- c(allcombs,matrixtolist(tmp,rc=2))
   }
   allcombs
}

# extracts rows (rc=1) or columns (rc=2) of a matrix,
# producing a list
matrixtolist <- function(rc,m) {
   if (rc == 1) {
      Map(function(rownum) m[rownum,],1:nrow(m))
   } else Map(function(colnum) m[,colnum],1:ncol(m))
}

# test data
gendata <- function(n,p) {
   x <<- matrix(rnorm(n*p),ncol=p)
   y <<- x%*%c(rep(0.5,p)) + rnorm(n)
}
```

As noted, the changes from the **snow** version are pretty small. References to clusters are gone, and we no longer export functions like **do1pset1()** to the workers, again because the workers already have them! The calls to **clusterApply()** have been replaced by **mclapply()**.[4]

Let's look at the calls to **mclapply()**:

```
out <- mclapply(tasks,dochunk2,x,y,xpx,xpy,allcombs,chunk,
   mc.cores=ncores,mc.preschedule=!dyn)
```

The call format (at least as used here) is almost identical to that of **clusterApply()**, with the main difference being that we specify the number of cores rather than specifying a cluster.

As with **snow**, **multicore** offers both static and dynamic scheduling, by setting the **mc.preschedule** parameter to either TRUE or FALSE, respectively. (The default is TRUE.) Thus here we simply set **mc.preschedule** to the opposite of **dyn**.

In that static case, **multicore** assigns loop iterations to the cores in a Round Robin manner as with **clusterApply()**.

For dynamic scheduling, **mclapply()** initially creates a number of R child processes equal to the specified number of cores; each one will handle one iteration. Then, whenever a child process returns its result to the original R process, the latter creates a new child, to handle another iteration.

Timings:

So, does it work well? Let's try it on a slightly larger problem than before—using eight cores again, same n and p, but with k = 5 instead of k = 3.

Here are the better times found in runs of the improved **snow** version we developed earlier:

```
> system.time(snowapr1(c8,x,y,5,dyn=TRUE,chunk=300))
   user  system elapsed
  7.561   0.368   8.398
> system.time(snowapr1(c8,x,y,5,dyn=TRUE,chunk=450))
   user  system elapsed
  5.420   0.228   7.175
> system.time(snowapr1(c8,x,y,5,dyn=TRUE,chunk=600))
   user  system elapsed
  3.696   0.124   6.677
```

[4]Though **mclapply()** still has **xpx** etc. as arguments, what will be copied will just be pointers to those variables in shared memory; no actual data will be copied. By contrast, if we run our previous **snow** code on clusters formed by **makeCluster()**, the data *will* be copies, via the sockets.

```
> system.time(snowapr1(c8,x,y,5,dyn=TRUE,chunk=800))
   user  system elapsed
  2.984   0.124   6.544
> system.time(snowapr1(c8,x,y,5,dyn=TRUE,chunk=1000))
   user  system elapsed
  2.505   0.092   6.441
> system.time(snowapr1(c8,x,y,5,dyn=TRUE,chunk=1200))
   user  system elapsed
  2.248   0.072   7.218
```

Compare to these results for the **multicore** version:

```
> system.time(mcapr(x,y,5,dyn=TRUE,chunk=50,ncores=8))
   user  system elapsed
 35.186  14.777   7.259
> system.time(mcapr(x,y,5,dyn=TRUE,chunk=75,ncores=8))
   user  system elapsed
 36.546  15.349   7.236
> system.time(mcapr(x,y,5,dyn=TRUE,chunk=100,ncores=8))
   user  system elapsed
 37.218   9.949   6.606
> system.time(mcapr(x,y,5,dyn=TRUE,chunk=125,ncores=8))
   user  system elapsed
 38.871   9.572   6.675
> system.time(mcapr(x,y,5,dyn=TRUE,chunk=150,ncores=8))
   user  system elapsed
 34.458   8.012   5.843
> system.time(mcapr(x,y,5,dyn=TRUE,chunk=175,ncores=8))
   user  system elapsed
 34.754   5.936   5.716
> system.time(mcapr(x,y,5,dyn=TRUE,chunk=200,ncores=8))
   user  system elapsed
 39.834   7.389   6.440
```

There are two points worth noting here. First, of course, we see that **multicore** did better, by about 10%. But also note that the **snow** version required much larger chunk sizes in order to do well. This should make sense, recalling the fact that the whole point of chunking is to amortize the overhead. Since the **snow** version has more overhead, it needs a larger chunk size to get good performance.

3.8 Issues with Chunk Size

We've seen here that program performance can be quite sensitive to the chunk size. If the chunk size is too small, we'll have more chunks to process, and thus will incur more overhead. But if the chunks are too large, we may have load balance problems near the end of the run.

So, how does one choose that value?

Data science is full of such vexing questions. Indeed, the example used earlier, in which we computed all possible regressions, was motivated by such a question: How do we choose the predictor set? That question has never been fully settled, despite a plethora of methods that have been developed. The situation for the chunk size is actually worse, since there are not even standard (if suboptimal) methods to deal with the problem.

In many applications, one must handle a sequence of problems, not just one. In such cases, one can determine a good chunk size via experimentation on the first one or two problems, and then use that chunk size from that point onward.

Note too that we have not tried the approach of using time-varying chunk size, mentioned briefly early in this chapter. Recall that the idea is to start out with large chunks for the early iterations, to reduce overhead, but then use smaller chunks near the end, to achieve better load balance.

You may wonder if this is even possible in **snow** or **multicore**. In fact, it is. Recall that we could achieve chunking with those two packages, even though neither offered chunking as an option; we simply had to code things properly.

Consider this simple example: We have 20 iterations and two processes. We could, say, define our chunks to consist of iterations 1-7, iterations 8-14, iterations 15-17 and iterations 18-20. In other words, we would have chunks of size 7, 7, 3 and 3.

Then we would make adjustments to the code accordingly.

So, we could indeed have time-varying chunk size, though at the expense of more complex coding. And there is no guarantee that the time-varying chunk size would give us much improvement, if any.

This issue will arise again in Section 5.3.

3.9 Example: Parallel Distance Computation

Say we have two data sets, with m and n observations, respectively. There are a number of applications in which we need to compute the mn pairs of distances between observations in one set and observations in the other. (The two data sets will be assumed separate from each other here, but the code could be adjusted if the sets are the same.)

Many clustering algorithms make use of distances, for example. These tend to be complex, so in order to have a more direct idea of why distances are important in many statistical applications, consider nonparametric regression.

Suppose we are predicting one variable from two others. For simplicity of illustration, let's use an example with concrete variables. Suppose we are predicting human weight from height and age. In essence, this involves expressing mean weight as a function of height and age, and then estimating the relationship from sample data in which all three variables are known, often called the *training set*. We also have another data set, consisting of people for whom only height and age are known, called the *prediction set*; this is used for comparing the performance of several models we ran on the training set, without the possible overfitting problem.

In nonparametric regression, the relationship between response and predictor variables is not assumed to have a linear or other parametric form. To guess the weight of someone in the prediction set, known to be 70 inches tall and 32 years old, we might look at people in our training set who are within, say, 2 inches of that height and 3 years of that age. We would then take the average weight of those people, and use it as our predicted weight for the 70-inch tall, age 32 person in our prediction set. As a refinement, we could give the people in our training sets who are very close to 70 inches tall and 32 years old more weight in this average.

Either way, we need to know the distances from observations in our training set to points in our prediction set. Suppose we have n people in our sample, and wish to predict p new people. That means we need to calculate np distances, exactly the setting described above. This could involve lots of computation, so let's see how we can parallelize it all, shown in the next section.

3.9.1 The Code

As usual, we hope to write parallel code that leverages existing R serial functions, in this case **pdist()**.

```
# finds distances between all possible pairs of rows
# in the matrix x and rows in the matrix y, as with
# pdist() but in parallel

# arguments:
#     cls:   cluster
#     x:   data matrix
```

```
#    y:  data matrix
#    dyn:  TRUE means dynamic scheduling
#    chunk:  chunk size
# return value:
#    full distance matrix, as pdist object

library(parallel)
library(pdist)

snowpdist <- function(cls,x,y,dyn=FALSE,chunk=1) {
   nx <- nrow(x)
   ichunks <- npart(nx,chunk)
   dists <-
      if (!dyn) { clusterApply(cls,ichunks,
         dochunk,x,y)
      } else clusterApplyLB(cls,ichunks,dochunk,x,y)
   tmp <- Reduce(c,dists)
   new("pdist", dist = tmp, n = nrow(x), p = nrow(y))
}

# process all rows in ichunk
dochunk <- function(ichunk,x,y) {
   pdist(x[ichunk,],y)@dist
}

# partition 1:m into chunks of approx. size chunk
npart <- function(m,chunk) {
   splitIndices(m,ceiling(m/chunk))
}
```

Let's see how this code works.

First, it builds upon the **pdist** package, available from R's CRAN repository of contributed code. The function **pdist()** in turn calls **Rpdist()**, written in C. Once again, we are heeding the advice in Section 1.1: In building our parallel code, we take advantage of powerful and efficiently implemented operations in R.

The basic approach is simple: We break the matrix **x** into chunks, then use **pdist()** to find the distances from rows in each chunk to **y**. However, we have some details to attend to in combining the results.

The **pdist** package defines an S4 class of the same name, the core of which is the distance matrix. Here is an example of such a matrix:

```
> x
     [,1] [,2]
[1,]    2    5
[2,]    4    3
> y
     [,1] [,2]
[1,]    1    4
[2,]    3    1
```

The distance matrix for these two data sets is

$$\begin{pmatrix} 1.414214 & 4.123106 \\ 3.162278 & 2.236068 \end{pmatrix} \tag{3.3}$$

The distance from row 1 of **x** to row 1 of **y** is $\sqrt{(1-2)^2 + (4-5)^2} = 1.414214$, while that from row 1 of **x** to row 2 of **y** is $\sqrt{(3-2)^2 + (1-5)^2} = 4.123106$. These numbers form row 1 of the distance matrix, and row 2 is formed similarly.

The function **pdist()** computes the distance matrix, returning it as the **dist** slot in an object of the class **pdist**:

```
> pdist(x,y)
An object of class "pdist"
Slot "dist":
[1] 1.414214 4.123106 3.162278 2.236068
attr(,"Csingle")
[1] TRUE

Slot "n":
[1] 2

Slot "p":
[1] 2

Slot ".S3Class":
[1] "pdist"
```

Note that the distance matrix is given as a one-dimensional vector, stringing all the rows together. You can convert it to a matrix if you wish:

```
> d <- pdist(x,y)
> as.matrix(d)
         [,1]     [,2]
[1,] 1.414214 4.123106
[2,] 3.162278 2.236068
```

With this in mind, look at the code:

```
    dists <-
       if (!dyn) { clusterApply(cls,ichunks,
          dochunk,x,y)
       } else clusterApplyLB(cls,ichunks,dochunk,x,y)
   tmp <- Reduce(c,dists)
   new("pdist", dist = tmp, n = nrow(x), p = nrow(y))
}
```

The list **dists** will contain the results of calling **pdist()** on the various chunks. Each one will be an object of class **pdist**. We need to essentially take them apart, combine the distance slots, then form a new object of class **pdist**.

Since the **dist** slot in a **pdist** object contains row-by-row distances anyway, we can simply use the standard R concatenate function **c()** to do the combining. We then use **new()** to create a grand **pdist** object for our final result.

If we simply wanted the distance matrix itself, we'd apply **as.matrix()** as the last step in **dochunk()**, and not call **new()** in **snowpdist()**.

3.9.2 Timings

As before, no attempt will be made here to do a general study of the efficiency of the code, but below are some sample timings, on 2 and 4 cores.

```
> genxy
function (n, k)
{
    x <<- matrix(runif(n * k), ncol = k)
    y <<- matrix(runif(n * k), ncol = k)
}
> genxy(15000,20)
> system.time(pdist(x,y))
```

```
   user  system elapsed
 40.459   6.144  46.885
> system.time(snowpdist(c2,x,y,chunk=500))
   user  system elapsed
 15.189   3.156  46.520
> system.time(snowpdist(c4,x,y,chunk=500))
   user  system elapsed
 15.749   3.620  34.537
```

The 2-node cluster failed to yield a speedup. The 4-node system was faster, but yielded a speedup of only about 1.36, rather than the theoretical value of 4.0.

Overhead seemed to have a major impact here, so a larger problem was investigated, with 50 variables instead of 20, and computing with up to 8 cores:

```
> genxy(15000,50)
> system.time(pdist(x,y))
   user  system elapsed
 88.925   5.597  94.901
> system.time(snowpdist(c2,x,y,chunk=500))
   user  system elapsed
 16.973   3.832  77.563
> system.time(snowpdist(c4,x,y,chunk=500))
   user  system elapsed
 17.069   3.800  49.824
> system.time(snowpdist(c8,x,y,chunk=500))
   user  system elapsed
 15.537   3.360  32.098
```

Here even use of only two nodes produced an improvement, and cluster sizes of 4 and 8 showed further speedups.

3.10 The foreach Package

Yet another popular R tool for parallelizing loops is the **foreach** package, available from the CRAN repository of contributed code. Actually **foreach** is more explicitly aimed at the loops case, as seen from its name, evoking **for** loops.

The package has the user set up a **for** loop, as in serial code, but then use the **foreach()** function instead of **for()**. One must also make one more small change, adding an operator, **%dopar%**, but that's all the user must do to parallelize his/her serial code.

Thus **foreach** has a very appealing simplicity. However, in some cases, this simplicity can mask major opportunities for achieving speedup, as will be seen in the example in the next section.

3.10.1 Example: Mutual Outlinks Problem

Here is **foreach** code for the mutual outlinks problem.

```
mutoutfe <- function(links) {
   require(foreach)
   nr <- nrow(links)
   nc <- ncol(links)
   tot = 0
   foreach(i = 1:(nr-1)) %dopar% {
      for (j in (i+1):nr) {
         for (k in 1:nc)
            tot <- tot + links[i,k] * links[j,k]
      }
   }
   tot / nr
}

simfe <- function(nr,nc,ncores) {
   require(doMC)  # loads 'parallel' too
   cls <- makeCluster(ncores)
   registerMC(cores=ncores)
   lnks <<- matrix(sample(0:1,(nr*nc),replace=TRUE),
      nrow=nr)
   print(system.time(mutoutfe(lnks)))
}
```

The function **mutoutfe()** above is an adaptation of the serial algorithm back in Chapter 1:

```
mutoutser <- function(links) {
   nr <- nrow(links)
   nc <- ncol(links)
   tot = 0
```

```
    for (i in 1:(nr-1)) {
        for (j in (i+1):nr) {
            for (k in 1:nc)
                tot <- tot + links[i,k] * links[j,k]
        }
    }
    tot / nr
}
```

The original **for** loop with index **i** has now been replaced by **foreach** and **%dopar%**:

```
foreach(i = 1:(nr-1)) %dopar% {
```

The user needs to also specify the platform to run on, the *backend* in **foreach** parlance. This can be **snow**, **multicore** or various other parallel software systems. This is the flexibility alluded to above—one can use the same code on different platforms.

To see how this works, here is a function that performs a speed test of the above code:

```
simfe <- function(nr,nc,ncores) {
    require(doMC)
    registerDoMC(cores=ncores)
    lnks <<- matrix(sample(0:1,(nr*nc),replace=TRUE),
        nrow=nr)
    print(system.time(mutoutfe(lnks)))
}
```

Here we've chosen to use the **multicore** backend. The package **doMC** is designed for this purpose. We call **registerDoMC()** to set up a call to **multicore** with the desired number of cores, and then when **foreach** within **mutoutfe()** runs, it uses that **multicore** platform.

Let's see how well it works:

```
> simfe(500,500,2)
   user  system elapsed
 17.392   0.036  17.663
> simfe(500,500,4)
   user  system elapsed
 52.900   0.176  13.578
> simfe(500,500,8)
   user  system elapsed
 62.488   0.352   7.408
```

3.10.2 A Caution When Using foreach

As noted, a strong appeal of **foreach** is that (for embarrassingly parallel problems) we can parallelize our serial code by simply changing just a single line in the latter. We just replace

```
for (i in irange)
```

by

```
foreach (i = irange) %dopar%
```

However, this simplicity can be quite deceiving in some cases.

For instance, the above timings for **foreach** on the mutual outlinks problem look good at first; the more cores we use, the shorter the run time. But something should trouble us here: We are checking one row at a time, i.e., one value of **i** at a time, and thus not taking advantage of R's fast matrix-multiplication capability, which gave us a dramatic increase in speed back in Section 1.4.5.

Indeed, the **snow** version, that did take advantage of matrix multiplication, is much faster here:

```
> simsnow
function(nr,nc,ncores) {
   require(parallel)
   lnks <<- matrix(sample(0:1,(nr*nc),replace=TRUE),
      nrow=nr)
   cls <- makeCluster(ncores)
   print(system.time(mutoutpar(cls)))
}
> simsnow(500,500,2)
   user  system elapsed
  0.272   0.076  11.266
> simsnow(500,500,4)
   user  system elapsed
  0.304   0.036   6.008
> simsnow(500,500,8)
   user  system elapsed
  0.348   0.040   3.407
```

Another example is our parallel distance computation in Section (3.9). Actually, you can see that this is a common scenario, occurring whenever there is an R function available that works most efficiently on chunks rather than on individual entities such as matrix rows.

The solution of course is easy: We simply incorporate chunking and matrix multiplication into the **foreach** version, and then have **i** range through the chunks accordingly. But the real moral of the story is that naive use of **foreach()** may mask real opportunities for speedup; we should not blithely assume that simply by making code parallel, we've achieved all the possible speedup.

This of course applies to **snow** and the other parallelization packages too, but again, we should be especially careful in the case of **foreach()** and avoid thinking that problems are easy to parallelize just because we use that package.

3.11 Stride

In discussions of parallel loop computation, one often sees the word *stride*, which refers to address spacing between successive memory accesses. Suppose we have a matrix having m rows and n columns. Consider the effects of, say, summing the elements in some column. If our storage uses column-major order, the accesses will be one word apart in memory, i.e. they will have stride 1. On the other hand, if we are using row-major storage, the stride will be n.

This is an issue in the context of memory bank structure (Section 2.5.1.1). Typically, *low-order interleaving* is used, meaning that consecutive words are stored in consecutive banks. If we have 4 banks, i.e. an *interleaving factor* of 4, then the first word will be stored in Bank 0, then next three in Banks 1, 2 and 3, then the next in Bank 0, and so on in a cyclical fashion.

The issue is avoiding *bank conflicts.* If say we have a stride of 1, then we can potentially keep all banks busy at once, the best we can hope for. Suppose on the other hand we have a stride of 4. This would be disastrous, as all accesses would go to the same bank, destroying our chance for parallel operation.

Even though we write our code at a high level, such as in R or C, it is important to keep in mind what stride will be implied by the algorithms we design. We may not know the bank interleaving factor of the machine our code will be run on, but at least such issues should be kept in mind. And in the case of GPU programming, truly maximizing speed may depend on this.

3.12 Another Scheduling Approach: Random Task Permutation

In situations in which nothing is known in advance about the iteration times, another possibility would be to randomize the order of the iterations before the computation begins.

For instance, consider the code in Section 3.4.2:

```
tasks <- if (!reverse) seq(1,ncombs,chunk) else
   seq(ncombs,1,-chunk)
nt <- length(tasks)
randpermut <- sample(1:nt,nt,replace=FALSE)
tasks <- tasks[randpermut]
if (!dyn) {
   out <- clusterApply(cls,tasks,dochunk,x,y,
      allcombs,chunk)
} else {
   out <- clusterApplyLB(cls,tasks,dochunk,x,y,
      allcombs,chunk)
```

3.12.1 The Math

If you are not interested in the mathematics, this subsection can easily be skipped, but it may provide insight for those who stay.

Say we have n iterations, with times $t_1, ..., t_n$, handled by p processes in static scheduling. Let π denote a random permutation of (1,...,n), and set

$$T_i = t_{\pi(i)}, \ i = 1, ...n \tag{3.4}$$

So the T_i are the randomly permuted t_i, thus random.

Then our ith process handles iterations π_s through π_e, where

$$s = (i-1)c + 1 \tag{3.5}$$

and

$$e = ic \tag{3.6}$$

with c being the chunk size:

$$c = n/p \tag{3.7}$$

(assuming n is divisible by p).

Let μ and σ^2 represent the mean and variance of the t_i:

$$\mu = \frac{1}{n} \sum_{i=1}^{n} t_i \tag{3.8}$$

$$\sigma^2 = \frac{1}{n} \sum_{i=1}^{n} (t_i - \mu)^2 \tag{3.9}$$

Note that these are not the mean and variance of some hypothesized parent distribution. No probabilistic model is being assumed for the t_i; indeed, they are not even being assumed random. So, μ and σ^2 are simply the mean and variance of the set of numbers $t_1, ..., t_n$.

Then $T_s, ..., T_e$ form a simple random sample (i.e. without replacement) from $t_1, ..., t_n$. From finite-population sampling theory, the total computation time U_i for the ith process has mean

$$c\mu \tag{3.10}$$

and variance

$$(1 - f)c\sigma^2 \tag{3.11}$$

where f = c/n.

The coefficient of variation of U_i, i.e. its standard deviation divided by its mean, is then

$$\frac{\sqrt{(1 - f)c\sigma^2}}{c\mu} \rightarrow 0 \text{ as } c \rightarrow \infty \tag{3.12}$$

Using standard analysis, say Chebyshev's Inequality, we know that a random variable with small coefficient of variation is essentially constant. Since $c = n/p$, then for large n, the T_i are essentially constant. (Here either p is assumed fixed, or p/n $\rightarrow$ 0.) In other words, *the Random method asymptotically achieves full load balance.*

Meanwhile, the Random method involves minimum possible scheduling overhead: A worker communicates only twice with the manager, once to receive data and once to return the results. In other words, the Random method is asymptotically optimal, in theory.

3.12.2 The Random Method vs. Others, in Practice

The intuition behind the Random method is that in large problems, the variance between processing time from thread to thread should be small. This implies good load balance.

Simulation results by the author have shown that the Random method generally performs fairly well. However, there are no "silver bullets" in the parallel processing world. Note the following:

- By randomizing the iteration ordering, one might lose some locality of reference, thus causing poor cache and/or virtual memory performance. This might be ameliorated by randomizing chunks instead of individual iterations.

- In the notation of the previous section, the theoretical justification for the Random method is based on the *variance* of the random variables T_i. Yet load balance involves the *maximum* of those random variables (say via the quantity maximum minus minimum), rather than their variance. For fixed n/p and increasing p, this could result in poor performance, as the chances of *some* process taking a long time for its chunk increase.

It is quite typical that either (a) the iteration times are known to be monotonic or (b) the overhead for running a task queue is small, relative to task times. In such cases, the Random method may not produce an improvement. However, it's something to keep in your loop scheduling toolkit.

3.13 Debugging snow and multicore Code

Generally debugging any code is hard, but it is extra difficult with parallel code. Just like a juggler, we have to be good at watching many things happening at once!

Worse, one cannot use debugging tools directly, such as R's built-in **debug()** and **browser()** functions. This is because our worker code is not running within a terminal/window environment. For the same reason, even calls to **print()** won't work.

So, let's see what we can do.

3.13.1 Debugging in snow

One can still use **browser()** in a kind of tricked-up way, which will be presented below. As it is a little clumsy, note that if you are using a Unix-family system (Mac/Linux/Cygwin/etc.), the **dbs()** function in my **partools** package (Section 3.5) automates the whole process for you! In that case, you can happily ignore the following.

Here is an outline of the procedure, say for a cluster of 2 workers:

- We insert **browser()** calls in the code to be executed at the workers.
- When we set up a cluster, we set **manual=TRUE** in our call to **makeCluster()**.
- That call will create the cluster, and then print out a message informing us at what IP address and port the manager is available.
- In 2 other windows on our screen, we start R, with an option to listen to commands from the manager at the given IP address.
- In each of the 2 worker windows, we instruct the worker to act on the commands sent by the manager.
- In the manager window, we call the code to be executed by the manager. That code will include a call to **clusterApply()**, or some other **snow** service. This causes the workers to start running our application!
- The workers will hit the **browser()** call, and we can then debug as usual in the two windows.

But again, all this is automated in **dbs()**.

3.13.2 Debugging in multicore

Unfortunately, the above scheme doesn't work for **multicore**.

One way around not having **print()** available is to use **cat()** and print to a file. Say we are trying to confirm that a certain variable **x** has the value 8, which we believe it should if our code is working right. (I call this The Principle of Confirmation, a fundamental rule in debugging: Step through the code, checking to see at various points whether the variables have the values we think they ought to have. Eventually we encounter a place that doesn't confirm, giving us a big clue as to the approximate location of the bug.) We could insert code like

```
cat("x is ",x,"\n",file="dbg")
```

If we next want to check a variable **y** we insert code like

```
cat("y is ",y,"\n",file="dbg",append=TRUE)
```

Note the **append** parameter. We can then inspect the file **dbg** from another window.

One major drawback of this is that all the debugging output from the various workers will be mixed together! This makes it hard to read, even if one also prints an ID number for each worker. This too is remedied by the **partools** package, via the function **dbsmsg()**, which has different workers write to different files; this function is platform-independent.

Chapter 4

The Shared-Memory Paradigm: A Gentle Introduction via R

The familiar model for the shared memory paradigm (in the hardware sense) is the multicore machine. The several parallel processes communicate with each other by accessing memory (RAM) cells that they have in common within a machine. This contrasts with *message-passing* hardware, in which there are a number of separate, independent machines, with processes communicating via a network that connects the machines.

Shared-memory programming is considered by many in the parallel processing community as being the clearest of the various paradigms available. Since programming development time is often just as important as program run time, the clear, concise form of the shared-memory paradigm can be a major advantage.

Another type of shared-memory hardware is accelerator chips, notably graphics processing units (GPUs). Here one can use one's computer's graphics card not for graphics, but for fast parallel computation of non-graphics specific operations, say matrix multiply.

Shared memory programming will be presented in three chapters. This chapter will present an overview of the subject, and illustrate it with the R package **Rdsm**. Though to get the most advantage from shared memory, one should program in C/C++, **Rdsm** enables one to achieve shared memory parallelism at the R level, which is much easier to program than C/C++.[1] The situation is analogous to the famous message-passing package MPI; to really exploit MPI's power, one should write in C/C++, but writing in R in **Rmpi**, an interface to MPI (Chapter 8), is much easier and is often "fast enough" (Section 1.1.1). Also, **Rdsm** shows that shared-memory programming can run significantly faster than other parallel packages for R, for some applications.

In addition to **Rdsm**'s direct value as a parallel package for R, it is also useful for us here in this chapter as a gentle introduction to shared-memory programming. The fact that R does the heavy lifting in terms of data and statistical operations means we can focus on learning shared-memory coding, more clearly than if we began with C/C++.

The chapter following this one will discuss shared-memory programming in C/C++, and the third chapter in the set will discuss GPU programming.

4.1 So, What Is Actually Shared?

The term *shared memory* means that the processors all share a common memory address space. Let's see what that really means.

4.1.1 Global Variables

We won't deal with machine language in this book, but a quick example will be helpful. A processor will typically include several *registers*, which are like memory cells but located inside the processor. In Intel processors, one of the registers is named RAX.[2] Note that on a multicore machine, each core will have its own registers, so that for example each core will have its own independent register named RAX.

Recall from Section 2.5.1.1 that the standard method of programming multicore machines is to set up *threads*. These are several instances of the same program running simultaneously, with the key feature that they share memory. To see what this means, suppose all the cores are running threads from our program, and that the latter includes the Intel machine language instruction

[1] One can also use FORTRAN, but its usage is much less common in data science.

[2] Some architectures are not register-oriented, but for simplicity we will assume a register orientation here.

```
movl 200, %eax
```

which copies the contents of memory location 200 to the core's RAX register.

Before continuing, it is worth asking the question, "Where did that 200 come from?" In our high-level language source code, say C, we may have had a variable named **z**. The C compiler would have decided at which memory address to store **z**, say 200, and may have translated one of our lines of C code accessing **z** to the above machine instruction.

The same principles hold for interpreted languages with virtual machines like R. If we have a variable **w** in our R code, the R interpreter will choose a memory location for it, and we will in the end be executing machine instructions like the one above.

Now, what happens when that machine instruction executes? Remember, there is only one memory location 200, shared by all cores, but each core has its own separate register set. If core 1 and core 4 happen to execute this same instruction at about the same time, the contents of memory location 200 will be copied to both core 1's RAX and core 4's RAX in the above example.

One technical issue that should be mentioned is that most machines today use *virtual addressing*, as explained in Chapter 2. Location 200 is actually mapped by the hardware to a different address, say 5208, during execution. But since in our example the cores are running threads from the same program, the virtual address 200 will map to that same location 5208, for all of the cores. Thus whether one is talking about a virtual or physical address, the key point is that all the cores are accessing the same actual memory cell.

4.1.2 Local Variables: Stack Structures

A subtlety here is that in referring to shared variables, we are typically talking about global variables, not local variables declared within a function. The locals are stored in shared memory too, but they are typically in *stacks*, which are sections of memory allocated to the threads, with a separate stack for each thread.

Items in stacks are in most machines referenced via a register called a *stack pointer*, which we will call SP here. Convention is that the stack grows toward memory address 0, so expansion of the stack by one word on a 64-bit (i.e., 8 byte) machine is accomplished by subtracting 8 from SP.

To illustrate this, consider the code

```
f <- function(x) {
    ...
    y <- 2
    z <- x + y
    ...
}

...
g <- function() {
    t <- 3
    v <- 6
    w <- f(v)
    u <- w + 1
}
```

What happens when the thread executes the call **f(v)** from with **g()**? The internal R code (or the compiled machine code, in the case of C/C++) will subtract 8 from SP, and then write 6 to the top of the (newly expanded) stack, i.e., to the word currently pointed to by SP. Then to start execution of **f()**, the code will do two things: (a) It will again decrement SP by 8, and write to the stack the address of the code following the call, in this case the code u <−w + 2 (to know where to come back later). (b) It will jump to the area of memory where **f()** is stored, thus starting execution of that function. The internal R code will then make space on the stack for **y**, again by subtracting 8 from SP, and when y <−2 is executed, 2 will be written to **y**'s location in the stack. Any code making use of **x** will be able to fetch it from within the stack, as it was placed there earlier.

The above is useful information in general, and will be discussed again later in the book, but for our purposes right now, the point is this: Since each core has a separate stack pointer, the stacks for the various threads will be in different sections of memory. Thus my threads-for-R package **Rdsm**, to be presented shortly, has been designed so that the local variable **y** will have a separate, independent instantiation at each thread. An **Rdsm** shared variable, by contrast, will have just one instantiation, readable and writable by all threads, as we'll see.

By the way, note too what happens when the function **f()** finishes execution and returns. The internal code will clean up the stack, by moving SP back to where it had been before writing the 6 and 2 to the stack. SP will now point to a word that contains the address recorded in step (a) above, and the code will now make a jump to that address, i.e., u <−w + 1 will be executed, exactly what we want. So, **g()** resumes execution, and its own local variables, such as **t**, are still available, as they are still on the stack, from the time **g()** itself had been called.

4.1.3 Non-Shared Memory Systems

In non-shared-memory systems, say a network of workstations on which we are running **Rmpi** or **snow**,[3] each workstation has its own memory, and each one will then have its own location 200, completely independent of the locations 200 at the other workstations' memories. Note, though, that each workstation might be running multicore hardware, in which case we have a hybrid system.

Note too that we can still run message-passing software such as **Rmpi** and **snow** on a multicore machine (and indeed, did so in earlier chapters). But in this case we simply are not taking advantage of the shared memory.[4] If we are using **Rmpi**, for instance, our several processes will not be threads, and virtual location 200 might map to 5208 on one core but 28888 on another.

4.2 Clarity of Shared-Memory Code

The shared-memory programming world view is considered by many in the parallel processing community to be one of the clearest forms of parallel programming.[5] Let's see why.

Suppose for instance we wish to copy **x** to **y**. In a message-passing setting such as **Rmpi**, **x** and **y** may reside in processes at network nodes 2 and 5, say. The programmer might write code to run on process 2 like

```
mpi.send.Robj(x,tag=0,dest=5)  # send to worker 5
```

and write code

[3]Recall that we use **snow** to refer to the portion of R's **parallel** package that originated as a package named **snow**.

[4]You may recall that if we create a **snow** cluster using **makeForkCluster()**; our globals are initially shared among the workers, but changes made by the workers to the globals won't be shared.

[5]See Chandra, Rohit (2001), *Parallel Programming in OpenMP*, Kaufmann, pp.10ff (especially Table 1.1), and Hess, Matthias *et al* (2003), Experiences Using OpenMP Based on Compiler Directive Software DSM on a PC Cluster, in *OpenMP Shared Memory Parallel Programming: International Workshop on OpenMP Applications and Tools*, Michael Voss (ed.), Springer, p.216.

```
# receive from worker 2
y <- mpi.recv.Robj(tag=0,source=2)
```

to run on process 5. By contrast, in a shared-memory environment, the variables **x** and **y** would be shared, and the programmer would merely write

```
y <- x
```

What a difference! Now that **x** and **y** are shared by the processes, we can access them directly, making our code vastly simpler.

Note carefully that we are talking about human efficiency here, not machine efficiency. Use of shared memory can greatly simplify our code, with far less clutter, so that we can write and debug our program much faster than we could in a message-passing environment. That doesn't necessarily mean our program itself has faster execution speed. We may have cache performance issues, for instance; we'll return to this point later.

It will turn out, though, that **Rdsm** can indeed enjoy a speed advantage over other parallel R packages for some applications. We'll return to this issue in Section 4.5.

4.3 High-Level Introduction to Shared-Memory Programming: Rdsm Package

Though one sometimes needs to write directly in C/C++ in order to truly maximize speed, it is highly desirable to stay within R whenever possible, in order to leverage R's powerful data manipulation and statistical operations. This is the philosophy underlying R packages such as **Rmpi** and **snow**.

However, those are message-passing approaches, and as mentioned above, the inherent simplicity of the shared-memory programming paradigm makes it highly desirable to get the best of both worlds—working in shared memory but writing in R. At the time of this writing, my package **Rdsm** is the only such package. You can download it from the R contributed package repository, CRAN.

R itself is not threaded (or more accurately, R does not make threading available at the R programming level). But **Rdsm** brings threads programming to R. And as noted earlier, in addition to **Rdsm**'s direct value as a parallel package for R, it is also useful for us here in this chapter as a gentle introduction to shared-memory programming.

4.3.1 Use of Shared Memory

As with **snow** and **Rmpi**, in **Rdsm** each process is a separate, independent instantiation of R. However, the difference is that with **Rdsm**, the processes must run on the same machine, and they share variables, in fact doing so via physical shared memory.

Modern operating systems allow the programmer to request that a chunk of memory be made available on a shared basis by any process that holds a certain code, a *key*. The **bigmemory** package in R's CRAN code repository enables this for R programmers, and **Rdsm** builds on this.[6]

Ironically, the shared-memory package **Rdsm** also uses the message-passing software **snow** for some infrastructure. Specifically, the **Rdsm** programmer makes a certain call to set up each shared variable, and **snow** is used to distribute the associated keys to the **Rdsm** threads, thus enabling the threads to share variables!

The shared variables must take the form of matrices, a **bigmemory** constraint. Of course, one can still have a shared scalar, as a 1×1 matrix. A shared matrix will have the class **"big.matrix"**.

Note that one must use brackets in referencing the shared matrices. For instance, to print the shared matrix **m**, write

```
print(m[ ,])
```

rather than

```
print(m)
```

The latter just prints out the location of the shared memory object.

4.4 Example: Matrix Multiplication

The standard "Hello World" example of the parallel processing community is matrix multiplication. Here is the **Rdsm** code, along with a small test.

[6] Though **Rdsm** is intended to run on shared-memory machines, **bigmemory** also enables shared variables with storage in shared disk files. Thus potentially, **Rdsm** could also be used to provide the shared-memory world view on a distributed system, e.g. clusters. However, this would require use of a shared file system that does appropriate updating, which is not the case for those in common usage, such as the Network File System (NFS). Thus as of this writing, **Rdsm** offers this only as an experimental feature.

4.4.1 The Code

```
# matrix multiplication; the product u %*% v is
# computed on the snow cluster cls, and written
# in-place in w; w is a big.matrix object

mmulthread <- function(u,v,w) {
   require(parallel)
   # determine which rows this thread will handle
   myidxs <-
      splitIndices(nrow(u),
         myinfo$nwrkrs)[[myinfo$id]]
   # compute this thread's portion of the product
   w[myidxs,] <- u[myidxs,] %*% v[,]
   0  # don't do expensive return of result
}

# test on snow cluster cls
test <- function(cls) {
   # init Rdsm
   mgrinit(cls)
   # set up shared variables a,b,c,
   mgrmakevar(cls,"a",6,2)
   mgrmakevar(cls,"b",2,6)
   mgrmakevar(cls,"c",6,6)
   # fill in some test data
   a[,] <- 1:12
   b[,] <- rep(1,12)
   # give the threads the function to be run
   clusterExport(cls,"mmulthread")
   # run it
   clusterEvalQ(cls,mmulthread(a,b,c))
   print(c[,])  # not print(c)!
}
```

Here is a test run:

```
> library(parallel)
> c2 <- makeCluster(2)  # 2 threads
> test(c2)
     [,1] [,2] [,3] [,4] [,5] [,6]
[1,]    8    8    8    8    8    8
[2,]   10   10   10   10   10   10
```

```
[3,]   12   12   12   12   12   12
[4,]   14   14   14   14   14   14
[5,]   16   16   16   16   16   16
[6,]   18   18   18   18   18   18
```

Here we first set up a two-node **snow** cluster **c2**. Remember, with **snow**, clusters are not necessarily physical clusters, and can be multicore machines. For **Rdsm** the latter is the case.

The code **test()** is run as the **snow** manager. It creates shared variables, then launches the **Rdsm** threads via **snow**'s **clusterEvalQ()**.

4.4.2 Analysis

The setup phase in **Rdsm** here involves the following.

First, **Rdsm**'s **mgrinit()** is called to initialize the **Rdsm** system, after which we use the **Rdsm** function **mgrmakevar()** to create three matrices in shared memory, **a**, **b** and **c** (**a** and **b** could have been simple R globals, rather than **Rdsm** variables). This action will distribute the necessary keys and the sizes of the shared objects to the **snow** worker nodes, i.e., the **Rdsm**. threads.

Then **snow**'s **clusterEvalQ()** is used to launch the threads. This function is the analog of R's **evalq()**, but it is run on the worker nodes, using their environments. On a quadcore machine running four **Rdsm** threads, for example, the above call

```
clusterEvalQ(cls,mmulthread(a,b,c))
```

will cause

```
mmulthread(a,b,c)
```

to run on all threads at once (though it probably won't be the case that all threads are running the same line of code simultaneously). Note that we first needed to ship the function **mmulthread()** itself to the threads, again because **clusterEvalQ()** runs our specified command in the context of the environments at the threads.

It is crucial to keep in mind the sharing, e.g. of **c[,]**. The manager acquires the key for a chunk of memory containing this variable and shares it with the workers, via **mgrmakevar()**. The workers write to that memory, and due to sharing—remember, sharing means they are all accessing the same physical memory locations—the manager can then read it and print out **c[,]**.

4.4.3 The Code

Now, how does **mmulthread()** work? The basic idea is break the rows of the argument matrix **u** into chunks, and have each thread work on one chunk.[7] Say there are 1000 rows, and we have a quadcore machine (on which we've set up a four-node **snow** cluster). Thread 1 would handle rows 1-250, thread 2 would work on rows 251-500 and so on.

The chunks are assigned in the code

```
myidxs <-
   splitIndices(nrow(u),myinfo$nwrkrs)[[myinfo$id]]
```

calling the **snow** function **splitIndices()**. For example, the value of **myidxs** at thread 2 will be 251:500. The built-in **Rdsm** variable **myinfo** is an R list containing **nwrkrs**, the total number of threads, and **id**, the ID number of the thread executing the above displayed line. On thread 2 in our example here, those numbers will be 4 and 2, respectively.

The reader should note the "me, my" point of view that is key to threads programming. Remember, each of the threads is (more or less) simultaneously executing **mmulthread()**. So, the code in that function must be written from the point of view of a particular thread. That's why we put the "my" in the variable name **myidxs**. We're writing the code from the anthropomorphic view of imagining ourselves as a particular thread executing the code. That thread is "me," and so the row indices are "my" indices, hence the name **myidxs**.

Each thread multiplies **v** by the thread's own chunk of **u**, placing the result in the corresponding chunk of **w**:

```
w[myidxs,] <- u[myidxs,] %*% v[,]
```

As noted in Section 4.2, unlike a message-passing approach, here we have no shipping of objects back and forth among threads; the objects are "already there," in shared memory, and we access them simply and directly.

Note in particular that the product matrix **w** is NOT part of the return value of the function. Instead, it is simply there in the matrix that the manager specified for **w** in the call to **mmulthread()**, in this case **c**. Hence in the code

```
clusterEvalQ(cls,mmulthread(a,b,c))
print(c[,])
```

[7]Some parallel algorithms partition both **u** and **v**. See Chapter 12.

we can simply print **c** to see the product of **a** and **b**.

4.4.4 A Closer Look at the Shared Nature of Our Data

As noted, the matrix **w** is not returned to the caller. Instead, it is simply available directly as a shared variable to all parties who hold the key for that variable.

Let's look at that a little more closely, running our test code through the debugger:

```
> debug(test)
> test(c2)
debugging in: test(c2)
...
debug at MM.tex#16: mgrmakevar(cls, "c", 6, 6)
Browse[2]> n
debug at MM.tex#17: a[, ] <- 1:12
Browse[2]> print(c)
An object of class "big.matrix"
Slot "address":
<pointer: 0x105804ce0>
```

As mentioned, **Rdsm** variables are **"big.matrix"** objects, an R S4 class. We see above that the **"big.matrix"** class consists primarily of a memory address, 0x105804ce0 in this case, which is the location of the actual shared matrix (and its associated information, such as the numbers of rows and columns).[8] Let's see who accesses that memory address:

The line

```
clusterEvalQ(cls,mmulthread(a,b,c))
```

executed by the manager, commands each worker to execute

```
mmulthread(a,b,c)
```

When they do so, the variable **c** in the call will be **w** within **mmulthread()**, and thus references to **w** will again be via that same address, 0x105804ce0.

[8] Readers who are well-versed in languages such as C may be interested in how the address is actually used. Basically, in R the array-access operations are themselves functions, such as the built-in function "[". As such, they can be overridden, as with operator overloading in C++, and **bigmemory** uses this approach to redirect expressions like w[2,5] to shared memory accesses. An earlier version of **Rdsm**, developed independently around the time **bigmemory** was being written, took the same approach.

As you can see, then, all of the threads are indeed sharing this matrix, as is the manager, since they are all accessing this spot in memory. So for example if any one of these entities writes to that shared object, the others will see the new values.

A side note: "Traditionally," R is a *functional language*, (mostly) free of *side effects*. To explain this concept, consider a function call **f(x)**. Any change that **f()** makes to **x** does not change the value of **x** in the caller. If it could change, this would be a *side effect* of the call, a commonplace occurrence in languages such as C/C++ but not in R. If we do want **x** to change in the caller, we must write **f()** to reurn the changed value of **x**, and then in the caller, reassign it, e.g.

```
x <- f(x)
```

As seen above, the **bigmemory** package, and thus **Rdsm**, do produce side effects.[9]

R has never been 100% free of side effects, e.g. due to use of the <<− operator, and the number of exceptions has been increasing. The **bigmemory** and **data.table** packages are examples, as is R's new reference classes. The motivation of allowing side effects is to avoid expensive copying of a large object when one changes only one small component of it. This is especially important for our parallel processing context; as mentioned earlier, needless copying of large objects can rob a parallel program of its speed.

The **Rdsm** package includes instructions for saving a key to a file and then loading it from another invocation of R on the same machine. The latter will then be able to access the shared variable as well. For example, one might write a Web crawler application, collecting Web data and storing it in shared member, and meanwhile monitor it interactively via a separate R process.

4.4.5 Timing Comparison

We won't do extensive timing experiments here, but let's just check that the code is indeed providing a speedup:

```
> n <- 5000
> m <- matrix(runif(n^2),ncol=n)
> system.time(m %*% m)
   user  system elapsed
```

[9]Indeed, this is one of **bigmemory**'s major user attractions, according to **bigmemory** coauthor Michael Kane.

```
345.077    0.220 346.356
> cls <- makeCluster(4)
> mgrinit(cls)
> mgrmakevar(cls,"msh",n,n)
> mgrmakevar(cls,"msh2",n,n)
> msh[,] <- m
> clusterExport(cls,"mmulthread")
> system.time(clusterEvalQ(cls,
     mmulthread(msh,msh,msh2)))
   user  system elapsed
  0.004   0.000  91.863
```

So, a four-fold increase in the number of cores yielded almost a fourfold increase in speed, very good.

4.4.6 Leveraging R

It was pointed out earlier that a good reason for avoiding C/C++ if possible is to be able to leverage R's powerful built-in operations. In this example, we made use of R's built-in matrix-multiply capability, in addition to its ability to extract subsets of matrices.

This is a common strategy. To solve a big problem, we break it into smaller ones of the same type, apply R's tools to the small problems, and then somehow combine to obtain the final result. This of course is a general parallel processing design pattern, not just for R, but with a difference in that here we need to find appropriate R tools. R is an interpreted language, thus with a tendency to be slow, but its basic operations typically make use of functions that are written in C, which are fast. Matrix multiplication is such an operation, so our approach here does work well.

4.5 Shared Memory Can Bring A Performance Advantage

In addition to the tendency of shared-memory code to be clearer and more concise, in many applications we can reap a significant performance gain as well. Message-passing systems by definition do a lot of copying of data, sometimes very large amounts of data, that is often unnecessary. With shared memory, we can read and write our needed data directly, as we saw earlier.

Note, though, that shared-memory access may involve hidden data copying. Each cache coherency transaction (Section 2.5.1.1) involves copying of data, and if such transactions occur frequently, it can add up to large amounts. Indeed, some of *that* copying may be unnecessary, say when a cache block is brought in but never used much afterward. Thus shared-memory programming is not necessarily a "win," but it will become clear below that it can be much faster for some applications, relative to other R parallel packages such as **snow**, **multicore**, **foreach** and even **Rmpi**.

To see why, here is a version of **mmulthread()** using the **snow** package:

```
snowmmul <- function(cls,u,v) {
   require(parallel)
   idxs <- splitIndices(nrow(u),length(cls))
   mmulchunk <-
      function(idxchunk) u[idxchunk,] %*% v
   res <- clusterApply(cls,idxs,mmulchunk)
   Reduce(rbind,res)
}
```

This test code was used:

```
testcmp <- function(cls,n) {
   require(Rdsm)
   require(parallel)
   mgrinit(cls)
   mgrmakevar(cls,"a",n,n)
   mgrmakevar(cls,"c",n,n)
   amat <- matrix(runif(n^2),ncol=n)
   a[,] <- amat
   clusterExport(cls,"mmulthread")
   print(system.time(clusterEvalQ(cls,
      mmulthread(a,a,c))))
   print(system.time(cmat <-
      nsnowmmul(cls,amat,amat)))
}
```

It turns out that **snow** is considerably slower than the **Rdsm** implementation, as seen in Table 4.1. The results are for various sizes of $n \times n$ matrices, and various numbers of cores. The machine had 16 cores, with a hyperthreading degree of 2 (Section 1.4.5.2).

One of the culprits is the line

```
Reduce(rbind,res)
```

n	# cores	**Rdsm** time	**Snow** time
2000	8	2.604	4.754
3000	16	9.280	13.187
3000	24	6.660	17.390

Table 4.1: Rdsm vs. snow

in the **snow** version. This involves a lot of copying of data, and possibly worse, multiple allocation of large matrices, greatly sapping speed. This is in stark contrast to the **Rdsm** case, in which the threads directly write their chunked-multiplication results to the desired output matrix. Note that the **Reduce()** operation itself is done serially, and though we might try to parallelize that too, that itself would require lots of copying, and thus may be difficult to make work well.

This of course was not a problem particular to **snow**. The same **Reduce()** operation or equivalent would be needed with **multicore**, **foreach** (using the **.combine** option), **Rmpi** and so on.[10] **Rdsm**, by writing the results directly to the desired output, avoids that problem.

It is clear that there are many applications with similar situations, in which tools like **snow** etc. do a lot of serial data manipulation following the parallel phase. In addition, iterative algorithms, such as k-means clustering (Section 4.9) involve repeated alternating between a serial and parallel phase. **Rdsm** should typically give faster speed than do the others in these applications.

On the other hand, some improvement can be obtained by using **unlist()** instead of **Reduce()**, writing the last line of **mmulthread()** as[11]

```
matrix(unlist(res),ncol=ncol(v))
```

Using this approach, the **snow** times for 16- and 24-node clusters on a 3000×3000 matrix seen above were reduced to 11.600 and 13.792, respectively (and were confirmed in subsequent runs not shown here).

The shared-memory vs. message-passing debate is a long-running one in the parallel processing community. It has been traditional to argue that the

[10] With **multicore**, we would have a little less copying, as explained in Section 3.7.1.
[11] As suggested by M. Hannon.

shared-memory paradigm doesn't scale well (Section 2.9), but the advent of modern multicore systems, especially GPUs, has done much to counter that argument.

4.6 Locks and Barriers

These are two central concepts in shared-memory programming. To explain them, we begin with the concept of *race conditions.*

4.6.1 Race Conditions and Critical Sections

Consider software to manage online airline reservations, and for simplicity, assume there is no overbooking of seats. At some point in the program, there will be a section consisting of one or more lines of code whose purpose is to perform the actual reservation of a seat. The customer's name and other data are entered into the database for the given flight on the given day. That section of code is known as a *critical section*, for the following reason.

Imagine a scenario in which two customers who want the given flight on the given day log in to the reservation system at about the same time. Each of them will be running a separate thread of the program (though of course they won't be aware of this). Suppose only one seat is left on the flight. It could happen that each thread finds that there is a seat remaining on the flight, and thus each thread enters the critical section—and thus each thread books its customer for the flight! One of the threads will be slightly ahead of the other, and the later thread will overwrite what the earlier one wrote. In other words, the first customer thinks she has successfully booked the flight, but actually has not.

Now you can see why such a section of code is called "critical." It is fraught with danger, with the situation being known as a *race condition.* (Sorry, you will be bombarded with terminology in the next few paragraphs.)

We say that the problem with the flight reservations above stemmed from a failure to update the reservation records *atomically.* The Greek word *atom* means "indivisible," and the allusion here is that trouble may arise if we "divide" i.e., separate, the read (checking for availability of a seat) and write (committing the seat to the customer) phases in the critical section, as opposed to doing both phases in one indivisible action. Doing that atomically would mean that a thread does the read and write as an indivisible pair, without having any other thread being able to act between the two phases, thus eliminating the danger.

4.6.2 Locks

What we need to avoid race conditions is a mechanism that will limit access to the critical section to only one thread at a time, known as *mutual exclusion.* A common mechanism is a *lock variable* or *mutex.* Most thread systems include functions **lock()** and **unlock()**, applied to a lock variable. Just before a critical section, one inserts a call to **lock()**, and we follow the section with a call to **unlock()**. Execution will work as follows.

Suppose the lock variable is already locked, due to some other thread currently being inside the critical section. Then the thread making the call to **lock()** will *block*, meaning that it will just freeze up for the time being, not returning yet. When the thread currently in the critical section finally exits, it will call **unlock()**, and the blocked thread will now unblock: This thread will enter the critical section, and relock the lock. (Of course, if several threads had been waiting at the lock, only one will succeed, and the others will continue waiting.)

To make this concrete, consider this toy example, in **Rdsm**. We've initialized **Rdsm** as a two-thread system, **c2**, and set up a 1×1 shared variable **tot**. The code simply repeatedly adds 1 to the total, **n** times, and thus should have a final value of **n**.

```
# this function is not reliable; if 2 threads both try
# to increment the total at about the same time, they
# could interfere with each other
s <- function(n) {
   for (i in 1:n) {
      tot[1,1] <- tot[1,1] + 1
   }
}

library(parallel)
c2 <- makeCluster(2)
clusterExport(c2,"s")
mgrinit(c2)
mgrmakevar(c2,"tot",1,1)
tot[1,1] <- 0
clusterEvalQ(c2,s(1000))
tot[1,1]  # should be 2000, but likely far from it
```

I did two runs of this. On the first one, the final value of **tot[1,1]** was 1021, while the second time it was 1017. Neither time did it come out 2000 as it "should." Moreover, the result was random.

The problem here is that the action

```
tot[1,1] <- tot[1,1] + 1
```

is not atomic. We could have the following sequence of events:

```
thread 1 reads tot[1,1], finds it to be 227
thread 2 reads tot[1,1], finds it to be 227
thread 1 writes 228 to tot[1,1]
thread 2 writes 228 to tot[1,1]
```

Here, **tot[1,1]** should be 229, but is only 228. No wonder in the experiments above, the total turned out to fall far short of the correct number, 2000.

But with locks, everything works fine. Continuing the above example, we run the code

```
# here is the reliable version, surrounding the
# increment by lock and unlock, so only 1 thread
# can execute it at once
s1 <- function(n) {
   for (i in 1:n) {
      rdsmlock("totlock")
      tot[1,1] <- tot[1,1] + 1
      rdsmunlock("totlock")
   }
}

mgrmakelock(c2,"totlock")
tot[1,1] <- 0
clusterExport(c2,"s1")
clusterEvalQ(c2,s1(1000))
tot[1,1]  # will print out 2000, the correct number
```

Here we call the **Rdsm** function **mgrmakelock()** to create a lock variable (we need to name it, as we may have several lock variables in a program), and then call **Rdsm**'s lock and unlock functions before and after adding 1 to the current total. Those latter two calls render the add-1-to-total operation atomic, and resulting code works properly.

4.6.3 Barriers

Another key structure is that of a *barrier*, which is used to synchronize all the threads. Suppose for instance that we need one thread to perform some special action, but that we need to have the other threads wait for that action to be performed. The threads system will provide a function to call that accomplishes this. In **Rdsm**, this function is named **barr()**, and when a thread calls it, the thread will block until all threads have called it. Afterward, they all proceed to the next line of code.

Note that internally a barrier needs to be implemented with a lock. You, the application programmer, won't see the lock (unless you're curious), but you do need to be aware that it is there, as locks adversely impact performance.

4.7 Example: Maximal Burst in a Time Series

Consider a time series of length n. We may be interested in bursts, periods in which a high average value is sustained. We might stipulate that we look only at periods of length k consecutive points, for a user-specified k. So, we wish to find the period of length k that has the maximal mean value.

4.7.1 The Code

Once again, let's leverage the power of R. The **zoo** time series package includes a function **rollmean(w,m)**, which returns all the means of blocks of length k, i.e., what are usually called *moving averages*—just what we need.

Here is the code:

```
# Rdsm code to find max burst in a time series;

# arguments:

#     x:   data vector
#     k:   block size
#     mas:   scratch space, shared, 1 x (length(x)-1)
#     rslts:  2-tuple showing the maximum burst value,
#             and where it starts; shared, 1 x 2
```

```
maxburst <- function(x,k,mas,rslts) {
   require(Rdsm)
   require(zoo)
   # determine this thread's chunk of x
   n <- length(x)
   myidxs <- getidxs(n-k+1)
   myfirst <- myidxs[1]
   mylast <- myidxs[length(myidxs)]
   mas[1,myfirst:mylast] <-
      rollmean(x[myfirst:(mylast+k-1)],k)
   # make sure all threads have written to mas
   barr()
   # one thread must do wrapup, say thread 1
   if (myinfo$id == 1) {
      rslts[1,1] <- which.max(mas[,])
      rslts[1,2] <- mas[1,rslts[1,1]]
   }
}

test <- function(cls) {
   require(Rdsm)
   mgrinit(cls)
   mgrmakevar(cls,"mas",1,9)
   mgrmakevar(cls,"rslts",1,2)
   x <<- c(5,7,6,20,4,14,11,12,15,17)
   clusterExport(cls,"maxburst")
   clusterExport(cls,"x")
   clusterEvalQ(cls,maxburst(x,2,mas,rslts))
   print(rslts[,])  # not print(rslts)!
}
```

The division of labor here involves assigning different chunks of the data to different **Rdsm** threads. To determine the chunks, we could call **snow**'s **splitIndices()** as before, but actually **Rdsm** provides a simpler wrapper for that, **getidxs()**, which we've called here, to determine where this thread's chunk begins and ends:

```
n <- length(x)
myidxs <- getidxs(n-k+1)
myfirst <- myidxs[1]
mylast <- myidxs[length(myidxs)]
```

We then call **rollmean()** on this thread's chunk, and write the results into this thread's section of **mas**:

```
mas[1,myfirst:mylast] <-
   rollmean(x[myfirst:(mylast+k-1)],k)
```

When all the threads are done executing the above line, we will be ready to combine the results. But how will we know when they're done? That's where the barrier comes in. We call **barr()** to make sure everyone is done, and then designate one thread to then combine the results found by the threads:

```
barr()  # make sure all threads have written to mas
if (myinfo$id == 1) {
   rslts[1,1] <- which.max(mas[,])
   rslts[1,2] <- mas[1,rslts[1,1]]
}
```

4.8 Example: Transforming an Adjacency Matrix

Here is another example of the use of barriers, this one more involved, both because the computation is a little more complex, and because we need *two* variables this time.

Say we have a graph with an adjacency matrix

$$\begin{pmatrix} 0 & 1 & 0 & 0 \\ 1 & 0 & 0 & 1 \\ 0 & 1 & 0 & 1 \\ 1 & 1 & 1 & 0 \end{pmatrix} \tag{4.1}$$

For example, the 1s in row 1, column 2 and row 4, column 1, signify that there is an edge from vertex 1 to vertex 2, and one from vertex 4 to vertex 1. We'd like to transform this to a two-column matrix that displays the links, in this case

$$\begin{pmatrix} 1 & 2 \\ 2 & 1 \\ 2 & 4 \\ 3 & 2 \\ 3 & 4 \\ 4 & 1 \\ 4 & 2 \\ 4 & 3 \end{pmatrix} \tag{4.2}$$

For instance, the (4,3) in the last row means there is an edge from vertex 4 to 3, corresponding to the 1 in row 4, column 3 of the adjacency matrix.

4.8.1 The Code

Here is **Rdsm** code for this:

```
# inputs a graph adjacency matrix, and outputs a
# two-column matrix listing the edges emanating from
# each vertex, each row of the form (fvert,tvert),
# i.e. "from vertex" and "to vertex"

# arguments:
#    adj:  adjacency matrix
#    lnks:  edges matrix; shared, nrow(adj)^2 rows
#           and 2 columns
#    counts:  numbers of edges found by each thread;
#              shared; 1 row, length(cls) columns
#              (i.e. 1 element per thread)

# in this version, the matrix lnks must be created
# prior to calling findlinks(); since the number of
# rows is unknown a priori, one must allow for the
# worst case, nrow(adj)^2 rows; after the run, the
# number of actual rows will be in
# counts[1,length(cls)], so that the excess
# remaining rows can be removed

findlinks <- function(adj,lnks,counts) {
   require(parallel)
   nr <- nrow(adj)
```

```
# get this thread's assigned portion of the
# rows of adj
myidxs <- getidxs(nr)

# determine where the 1s are in this thread's
# portion of adj; for each row number i in myidxs,
# an element of myout will record the column
# locations of the 1s in that row, i.e. record the
# edges out of vertex i
myout <- apply(adj[myidxs,],1,
   function(onerow) which(onerow==1))

# this thread will now form its portion of lnks,
# storing in tmp
tmp <- matrix(nrow=0,ncol=2)
my1strow <- myidxs[1]
for (idx in myidxs)
   tmp <- rbind(tmp,convert1row(idx,
      myout[[idx-my1strow+1]]))

# we need to know where in lnks to put tmp; e.g.
# if threads 1 and 2 find 12 and 5 edges, then
# thread 3's portion of lnks will begin at row
# 12+5+1 = 18 of lnks

# so, let's find cumulative edge sums, and
# place them in counts
nmyedges <-
   Reduce(sum,lapply(myout,length))  # my count
me <- myinfo$id
counts[1,me] <- nmyedges
barr()  # wait for all threads to write to counts

# determine where in lnks the portion of thread
# 1 ends; thread 2's portion of lnks begins
# immediately after thread 1's, etc., so we need
# cumulative sums, which we'll place in counts;
# we'll have thread 1 perform this task, though
# any thread could do it
if (me == 1) counts[1,] <- cumsum(counts[1,])
barr()  # others wait for thread 1 to finish

# this thread now places tmp in its proper
```

```
   # position within lnks
   mystart <- if (me == 1) 1 else counts[1,me-1] + 1
   myend <- mystart + nmyedges - 1
   lnks[mystart:myend,] <- tmp

   0  # don't do expensive return of result
}

# if, say, row 5 in adj has 1s in columns 2, 3 and 8,
# this function returns the matrix
#     5 2
#     5 3
#     5 8
convert1row <- function(rownum,colswith1s) {
   if (is.null(colswith1s)) return(NULL)
   cbind(rownum,colswith1s)  # use recycling
}

test <- function(cls) {
   require(Rdsm)
   mgrinit(cls)
   mgrmakevar(cls,"x",6,6)
   mgrmakevar(cls,"lnks",36,2)
   mgrmakevar(cls,"counts",1,length(cls))
   x[,] <- matrix(sample(0:1,36,replace=TRUE),ncol=6)
   clusterExport(cls,"findlinks")
   clusterExport(cls,"convert1row")
   clusterEvalQ(cls,findlinks(x,lnks,counts))
   print(lnks[1:counts[1,length(cls)],])
}
```

The division of labor here involves assigning different chunks of rows of the adjacency matrix to different **Rdsm** threads. We first partition the rows, as before, then determine the locations of the 1s in this thread's chunk of rows:

```
myidxs <- getidxs(nr)
myout <- apply(a[myidxs,],1,function(rw) which(rw==1))
```

The R list **myout** will now give a row-by-row listing of the column numbers of all the 1s in the rows of this thread's chunk. Remember, our ultimate output matrix, **lnks**, will have one row for each such 1, so the information in **myout** will be quite useful.

Here is how it uses that information, for a given row:

```
convert1row <- function(rownum, colswith1s) {
   if (is.null(colswith1s)) return(NULL)
   cbind(rownum, colswith1s)  # use recycling
}
```

This function returns a chunk that will eventually go into **lnks**, specifically the chunk corresponding to row **rownum** in **adj**. The code to form all such chunks for our given thread is

```
tmp <- matrix(nrow=0,ncol=2)
my1strow <- myidxs[1]
for (idx in myidxs)  tmp <-
   rbind(tmp, convert1row(idx, myout[[idx-my1strow+1]]))
```

Note that here the code needed to recognize the fact that the information for row number **idx** in **adj** is stored in element **idx - my1strow + 1** of **myout**.

Now that this thread has computed its portion of **lnks**, it must place it there. But in order to do so, this thread must know where in **lnks** to start writing. And for that, this thread needs to know how many 1s were found by threads prior to it. If for instance thread 1 finds eight 1s and thread 2 finds three, then thread 3 must start writing at row $8 + 3 + 1 = 12$ in **lnks**. Thus we need to find the overall 1s counts (across all rows of a thread) for each thread,

```
nmyedges <- Reduce(sum, lapply(myout, length))
```

and then need to find cumulative sums, and share them. To do this, we'll have (for instance) thread 1 find those sums, and place them in our shared variable **counts**:

```
me <- myinfo$id
counts[1,me] <- nmyedges
barr()
if (me == 1) {
   counts[1,] <- cumsum(counts[1,])
}
barr()
```

Note the barrier calls just before and just after thread 1. The first call is needed because thread 1 can't start finding the cumulative sums before all the individual counts are ready. Then we need the second barrier, because all the threads will be making use of the cumulative sums, and we need to be sure those sums are ready first. These are typical examples of barrier use.

Now that our thread knows where in **lnks** to write its results, it can go ahead:

```
mystart <- if (me == 1) 1 else counts[1,me-1] + 1
myend <- mystart + nmyedges - 1
lnks[mystart:myend,] <- tmp
```

4.8.2 Overallocation of Memory

A problem above is having to allocate the **lnks** matrix to handle the worst case, thus wasting space and execution time. The problem is that we don't know in advance the size of our "output," in this case the argument **lnks**. In our little example above, the adjacency matrix was of size 4x4, while the edges matrix was 7x2. We know the number of columns in the edges matrix will be 2, but the number of rows is unknown *a priori*.

Note that the user can determine the number of "real" rows in **lnks** by inspecting **counts[1,length(cls)]** after the call returns, as seen in the test code. One could copy the "real" rows to another matrix, then deallocate the big one.

One alternate approach would be to postpone allocation until we know how big the **lnks** matrix needs to be, which we will know after the cumulative sums in **counts** are calculated. We could have thread 1 then create the shared matrix **lnks**, by calling **bigmemory** directly rather than using **mgrmakevar()**. To distribute the shared-memory key for this matrix, thread 1 would save the **bigmemory** descriptor to a file, then have the other threads get access to **lnks** by loading from the file.

Actually, this problem is common in parallel processing applications. We will return to it in Section 5.4.2.

4.8.3 Timing Experiment

For comparison, here is a serial version of the code:

```
> getlinksnonpar
function(a,lnks) {
   nr <- nrow(a)
```

```
    myout <- apply(a[,],1,function(rw) which(rw==1))
    nmyedges <- Reduce(sum, lapply(myout, length))
    lnksidx <- 1
    for (idx in 1:nr) {
        jdx <- idx
        myoj <- myout[[jdx]]
        endwrite <- lnksidx + length(myoj) - 1
        if (!is.null(myoj)) {
            lnks[lnksidx:endwrite,] <- cbind(idx,myoj)
        }
        lnksidx <- endwrite + 1
    }
    0
}
```

```
> n <- 10000
> system.time(findlinks(x,lnks))
   user  system elapsed
 26.170   1.224  27.516
```

(For convenience, we are still using **Rdsm** to set up the shared variables, though we run in non-**Rdsm** code.)

Now try the parallel version:

```
> cls <- makeCluster(4)
> mgrinit(cls)
> mgrmakevar(cls,"counts",1,length(cls))
> mgrmakevar(cls,"x",n,n)
> mgrmakevar(cls,"lnks",n^2,2)
> x[,] <- matrix(sample(0:1,n^2,replace=TRUE),ncol=n)
> clusterExport(cls,"findlinks")
> clusterExport(cls,"convert1row")
> system.time(clusterEvalQ(cls,findlinks(x,lnks,counts)))
   user  system elapsed
  0.000   0.000   7.783
```

So, the parallel code did indeed speed things up.

4.9 Example: k-Means Clustering

In discussion of parallel computation for data science, an example application almost as common as matrix multiplication is k-means clustering. The goal is to form k groups from our data matrix, hopefully in a way that makes visual (or other) sense. Let's see how that can be implemented in **Rdsm**.

The general k-means method itself is quite simple, using an iterative algorithm. At any step during the iteration process, the k groups are summarized by their centroids.[12] We iterate the following:

1. For each data point, i.e., each row of our data matrix, determine which centroid this point is closest to.
2. Add this data point to the group corresponding to that centroid.
3. After all data points are processed in this manner, update the centroids to reflect the current group memberships.
4. Next iteration.

This example will bring in a concept in shared-memory work that didn't arise in our matrix multiplication example, related to the phrase, "After all data points are processed..." in step 3. Some other new concepts will come up as well, all to be explained below.

4.9.1 The Code

So, here is the code, again with a small test function:

```
# k-means clustering on the data matrix x, with
# k clusters and ni iterations; final cluster
# centroids placed in cntrds

# initial centroids taken to be k randomly chosen
# rows of x; if a cluster becomes empty, its new
# centroid will be a random row of x

library(Rdsm)
```

[12] If we have m variables, then the centroid of a group is the m-element vector of means of those variables within this group.

```
# arguments:
#    x:  data matrix x; shared
#    k:  number of clusters
#    ni:  number of iterations
#    cntrds:  centroids matrix; row i is centroid i;
#             shared, k by ncol(x)
#    cinit:  optional initial values for centroids;
#            k by ncol(x)
#    sums:  scratch matrix; sums[j,] contains count,
#           sum for cluster j; shared, k by 1+ncol(x)
#    lck:  lock variable; shared

kmeans <- function(x,k,ni,cntrds,sums,lck,cinit=NULL) {
   require(parallel)
   require(pdist)
   nx <- nrow(x)
   # get my assigned portion of x
   myidxs <- getidxs(nx)
   myx <- x[myidxs,]
   # random initial centroids if none specified
   if (is.null(cinit)) {
      if (myinfo$id == 1)
         cntrds[,] <- x[sample(1:nx,k,replace=FALSE),]
      barr()
   } else cntrds[,] <- cinit

   # mysum() sums the rows in myx corresponding to the
   # indices idxs; we also produce a count of those rows
   mysum <- function(idxs,myx) {
      c(length(idxs),colSums(myx[idxs,,drop=FALSE]))
   }
   for (i in 1:ni) {  # ni iterations
      # cluster node 1 is sometimes asked to do
      # some "housekeeping"
      if (myinfo$id == 1) {
         sums[] <- 0
      }
      # other nodes wait for node 1 to do its work
      barr()
      # find distances from my rows of x to the
      # centroids, then find which centroid is closest
      # to each such row
```

```
      dsts <-
         matrix(pdist(myx,cntrds[,])@dist,ncol=nrow(myx))
      nrst <- apply(dsts,2,which.min)
      # nrst[i] contains the index of the nearest
      # centroid to row i in myx
      tmp <- tapply(1:nrow(myx),nrst,mysum,myx)
      # in the above, we gather the observations
      # in myx whose closest centroid is centroid j,
      # and find their sum, placing it in # tmp[j];
      # the latter will also have the count of such
      # observations # in its leading component;
      # next, we need to add that to sums[j,],
      # as an atomic operation
      realrdsmlock(lck)
      # j values in tmp will be strings, so convert
      for (j in as.integer(names(tmp))) {
         sums[j,] <- sums[j,] + tmp[[j]]
      }
      realrdsmunlock(lck)
      barr()  # wait for sums[,] to be ready
      if (myinfo$id == 1) {
         # update centroids, using a random
         # data point if a cluster becomes empty
         for (j in 1:k) {
            # update centroid for cluster j
            if (sums[j,1] > 0) {
               cntrds[j,] <- sums[j,-1] / sums[j,1]
            } else cntrds[j] <<- x[sample(1:nx,1),]
         }
      }
   }
   0  # don't do expensive return of result
}

test <- function(cls) {
   library(parallel)
   mgrinit(cls)
   mgrmakevar(cls,"x",6,2)
   mgrmakevar(cls,"cntrds",2,2)
   mgrmakevar(cls,"sms",2,3)
   mgrmakelock(cls,"lck")
   x[,] <- matrix(sample(1:20,12),ncol=2)
   clusterExport(cls,"kmeans")
```

```
   clusterEvalQ(cls,kmeans(x,2,1,cntrds,sms,"lck",
      cinit=rbind(c(5,5),c(15,15))))
}

test1 <- function(cls) {
   mgrinit(cls)
   mgrmakevar(cls,"x",10000,3)
   mgrmakevar(cls,"cntrds",3,3)
   mgrmakevar(cls,"sms",3,4)
   mgrmakelock(cls,"lck")
   x[,] <- matrix(rnorm(30000),ncol=3)
   ri <- sample(1:10000,3000)
   x[ri,1] <- x[ri,1] + 5
   ri <- sample(1:10000,3000)
   x[ri,2] <- x[ri,2] + 5
   clusterExport(cls,"kmeans")
   clusterEvalQ(cls,kmeans(x,3,50,cntrds,sms,"lck"))
}
```

Let's first discuss the arguments of **kmeans()**. Our data matrix is **x**, which is described in the comments as a shared variable (on the assumption that it will often be such) but actually need not be.

By contrast, **cntrds** needs to be shared, as the threads repeatedly use it as the iterations progress. We have thread 1 writing to this variable,

```
if (myinfo$id == 1) {
   for (j in 1:k) {
      if (sums[j,1] > 0) {
         cntrds[j,] <<- sums[j,-1] / sums[j,1]
      } else cntrds[j] <<- x[sample(1:nx,1),]
   }
}
```

at the end of each iteration, and all threads reading it:

```
dsts <- matrix(pdist(myx,cntrds[,])@dist,
   ncol=nrow(myx))
```

If **cntrds** were not shared, the whole thing would fall apart. When thread 1 would write to it, it would become a local variable for that thread, and the new value would not become visible to the other threads. Note that as in our previous examples, we store our function's final result, in this case **cntrds**, in a shared variable, rather than as a return value.

The argument **sums** is also shared by necessity. It is only used to store intermediate results, but again this variable is written to by some threads and subsequently read by others, hence must be shared.

Another argument to **kmeans()** that is shared is **lck**, a lock variable, to be discussed below.

So, let's look at the actual code, starting with

```
# get my assigned portion of x
myidxs <- getidxs(nx)
myx <- x[myidxs,]
```

Once again our approach will be to break the data matrix into chunks of rows. Each thread will handle one chunk, finding distances from rows in its chunk to the current centroids. How is the above code preparing for this?

Note again the "me, my" point of view here, pointed out in Section 4.4 and present in almost any threads function. The code here is written from the point of view of a particular thread. So, the code first needs to determine this thread's rows chunk.

Why have this separate variable, **myx**? Why not just use **x[myidxs,]**? First, having the separate variable results in less cluttered code. But secondly, repeated access to **x** could cause a lot of costly cache misses and cache coherency actions.

Next we see another use of barriers:

```
if (is.null(cinit)) {
   if (myinfo$id == 1)
      cntrds[,] <- x[sample(1:nx,k,replace=FALSE),]
   barr()
} else cntrds[,] <- cinit
```

We've set things up so that if the user does not specify the initial values of the centroids, they will be set to k random rows of **x**. We've written the code so that thread 1 performs this task, but we need the other threads to wait until the task is done. If we didn't do that, one thread might race ahead and start accessing **cntrds** before it is ready. Our call to **barr()** ensures that this won't happen.

We have a similar use of a barrier at the beginning of the main loop:

```
if (myinfo$id == 1) {
   sums[] <- 0
}
```

```
barr()  # other nodes wait for node 1 to do its work
```

We need to compute the distances to the various centroids from all the rows in this thread's portion of our data:

```
dsts <-
   matrix(pdist(myx, cntrds[,])@dist, ncol=nrow(myx))
```

R's **pdist** package comes to the rescue! This package, which we saw in Section 3.9, finds all distances from the rows of one matrix to the rows of another, exactly what we need. So, here again, we are leveraging R! (Indeed, an alternate way to parallelize the computation from what we are doing here would be to parallelize **pdist()**, say using **Rdsm** instead of **snow** as before.)

Next, we leverage R's **which.min()** function, which finds indices of minima (not the minima themselves). We use this to determine the new group memberships for the data points in **myx**:

```
nrst <- apply(dsts,2,which.min)
```

Next, we need to collect the information in **nrst** into a more usable form, in which we have, for each centroid, a vector stating the indices of all rows in **myx** that now will belong to that centroid's group. For each centroid, we'll also need to sum all such rows, in preparation for later averaging them to find the new centroids.

Again, we can leverage R to do this quite compactly (albeit needing a bit of thought):

```
mysum <- function(idxs,myx) {
   c(length(idxs),colSums(myx[idxs,,drop=FALSE]))
}
...
tmp <- tapply(1:nrow(myx),nrst,mysum,myx)
```

But remember, *all* the threads are doing this! For instance, thread 1 is finding the sum of its rows that are now closest to centroid 6, but thread 4 is doing the same. For centroid 6, we will need the sum of all such rows, across all such threads.

In other words, multiple threads may be writing to the same row of **sums** at about the same time. Race condition ahead! So, we need a lock:

```
lock(lck)
for (j in names(tmp)) {
```

```
    j <- as.integer(j)
    sums[j,] <- sums[j,] + tmp[[j]]
}
unlock(lck)
```

The **for** loop here is a critical section. Without the restriction, chaos could result. Say for example two threads want to add 3 and 8 to a certain total, respectively, and that the current total is 29. What could happen is that they both see the 29, and compute 32 and 37, respectively, and then write those numbers back to the shared total. The result might be that the new total is either 32 or 37, when it actually should be 40. The locks prevent such a calamity.

A refinement would be to set up k locks, one for each row of **sums**. As noted earlier, locks sap performance, by temporarily serializing the execution of the threads. Having k locks instead of one might ameliorate the problem here.

After all the threads are done with this work, we can have thread 1 compute the new averages, i.e., the new centroids. But the key word in the last sentence is "after." We can't let thread 1 do that computation until we are sure that all the threads are done. This calls for using a barrier:

```
barr()
if (myinfo$id == 1) {
    for (j in 1:k) {
       if (sums[j,1] > 0) {
          cntrds[j,] <<- sums[j,-1] / sums[j,1]
       } else cntrds[j] <<- x[sample(1:nx,1),]
    }
}
```

As noted earlier, the shared variable **sums** serves as storage for intermediate results, not only sums of the data points in a group, but also their counts. We can now use that information to compute the new centroids:

```
if (myinfo$id == 1) {
    for (j in 1:k) {
       # update centroid for cluster j
       if (sums[j,1] > 0) {
          cntrds[j,] <- sums[j,-1] / sums[j,1]
       } else cntrds[j] <<- x[sample(1:nx,1),]
    }
}
```

4.9.2 Timing Experiment

Let n denote the number of rows in our data matrix. With k clusters, we have to compute nk distances per iteration, and then take n minima. So the time complexity is $O(nk)$.

This is not very promising for parallelization. In many cases $O(n)$ (fixing k here) does not provide enough computation to overcome overhead issues. However, with our code here, there really isn't much overhead. We copy the data matrix just once,

```
myx <- x[myidxs,]
```

and thus avoid problems of contention for shared memory and so on.

It appears that we can indeed get a speedup from our parallel version some cases:

```
> x <- matrix(runif(100000*25),ncol=25)
> system.time(kmeans(x,10))   # kmeans(), base R
   user  system elapsed
  8.972   0.056   9.051
> cls <- makeCluster(4)
> mgrinit(cls)
> mgrmakevar(cls,"cntrds",10,25)
> mgrmakevar(cls,"sms",10,26)
> clusterExport(cls,"kmeans")
> mgrmakevar(cls,"x",100000,25)
> x[,] <- x
> system.time(clusterEvalQ(cls,
     kmeans(x,10,10,cntrds,sms,lck)))
   user  system elapsed
  0.000   0.000   4.086
```

A bit more than 2X speedup for four cores, fairly good in view of the above considerations.

4.10 Further Reading

For more details on the issues raised with the question, "What actually is shared?", the nature of stacks, the implementation of locks and so on, see my open source book on computer systems, *An Introduction to Com-*

puter Systems, N. Matloff, `http://heather.cs.ucdavis.edu/~matloff/50/PLN/CompSystsBook.pdf`.

Barrier implementation is discussed in many books, such as *The Art of Concurrency*, by Clay Breshears, O'Reilly, 2009. That book also discusses "strategy" for writing parallel code, as does *Patterns for Parallel Programming*, by T. Mattson *et al*, Addison-Wesley, 2004.

Chapter 5

The Shared-Memory Paradigm: C Level

As mentioned in Section 1.1, an increasingly common usage of R is "R+X," in which coding is done in a combination of R and some other language. From the beginning of R, a common example has involved the C/C++ languages playing the role of X, and this is especially important for parallel computation.

The standard method for programming directly on multicore machines, is to use threads libraries, which are available for all modern operating systems. On Unix-family systems (Linux, Mac), for example, the **pthreads** library is quite popular.

The programmer calls functions in the threads library. For instance, one calls the **pthread_mutex_lock()** function in **pthreads** to lock a lock variable. However, this can become very tedious, so higher-level libraries were developed specifically with parallel computation in mind, such as OpenMP, Intel's Threads Building Blocks and Cilk++, all of them cross-platform software.[1] Though the latter two are very powerful, here we mainly cover OpenMP, the most popular of the three. We will use C as our language.

(Note to the reader: If you do not have background in C, you should still be able to follow the code here fairly well, on the strength of your knowledge of R programming. There is an introduction in Appendix C.)

[1] Currently OpenMP is not supported by **clang**, the default compiler for Macs. One thus needs to install another compiler, such as **gcc**.

5.1 OpenMP

An OpenMP application still uses threads, but at a higher level of abstraction. One accesses OpenMP through C, C++ or FORTRAN. R users can write an OpenMP application in one of those languages, and then call the application from R, using either the **.C()** or **.Call()** functions available in R for that purpose; if you do much of this, you should probably use the **Rcpp** package as your interface. To keep things simple, we will stick to the **.C()** interface here for the time being, and switch to **.Call()/Rcpp** after a while. (In order to facilitate interface with R, we use C's **double** type instead of **float**.)

5.2 Example: Finding the Maximal Burst in a Time Series

Consider a time series of length n, in the context of our example in Section 4.7, but with a modified goal, to find the period of at least k consecutive time points that has the maximal mean value.

Denote our time series by $x_1, x_2, ..., x_n$. Consider checking for bursts that begin at x_i. We could check for bursts of length $k, k+1, ..., n-i+1$, i.e., about $n-i-k$ different cases. Since i itself can take on $O(n)$ values, the time complexity of this application is, for fixed k and varying n, $O(n^2)$. This growth rate in n suggests that this is a good candidate for parallelization.

5.2.1 The Code

Here is the code, written without an R interface for the time being.

We will discuss it in detail below, but you should glance through it first. As you do, note the *pragma* lines, such as

```
#pragma omp single
```

These are actually OpenMP directives, which instruct the compiler to insert certain thread operations at that point.

For convenience, the code will assume that the time series values are non-negative.

```
// OpenMP example program, Burst.c; burst() finds
```

```
// period of highest burst of activity in a time series

#include <omp.h>
#include <stdio.h>
#include <stdlib.h>

// arguments for burst()

//    inputs:
//       x:  the time series, assumed nonnegative
//       nx:  length of x
//       k:  shortest period of interest
//    outputs:
//       startmax, endmax:  pointers to indices of
//       the maximal-burst period
//       maxval:  pointer to maximal burst value

// finds the mean of the block between y[s] and y[e]
double mean(double *y, int s, int e) {
   int i; double tot = 0;
   for (i = s; i <= e; i++) tot += y[i];
   return tot / (e - s + 1);
}

void burst(double *x, int nx, int k,
   int *startmax, int *endmax, double *maxval)
   {
   int nth;  // number of threads
   #pragma omp parallel
   {  int perstart, // period start
          perlen,  // period length
          perend,  // perlen end
          pl1;  // perlen - 1
      // best found by this thread so far
      int mystartmax, myendmax;  // locations
      double mymaxval;  // value
      // scratch variable
      double xbar;
      int me;  // ID for this thread
      #pragma omp single
      {
          nth = omp_get_num_threads();
      }
```

```
      me = omp_get_thread_num();
      mymaxval = -1;
      #pragma omp for
      for (perstart = 0; perstart <= nx-k;
            perstart++) {
         for (perlen = k; perlen <= nx - perstart;
               perlen++) {
            perend = perstart+perlen-1;
            if (perlen == k)
               xbar = mean(x,perstart,perend);
            else {
               // update the old mean
               pl1 = perlen - 1;
               xbar =
                  (pl1 * xbar + x[perend]) / perlen;
            }
            if (xbar > mymaxval) {
               mymaxval = xbar;
               mystartmax = perstart;
               myendmax = perend;
            }
         }
      }
      #pragma omp critical
      {
         if (mymaxval > *maxval) {
            *maxval = mymaxval;
            *startmax = mystartmax;
            *endmax = myendmax;
         }
      }
   }
}

// here's our test code

int main(int argc, char **argv)
{
   int startmax, endmax;
   double maxval;
   double *x;
   int k = atoi(argv[1]);
   int i,nx;
```

```
   nx = atoi(argv[2]);  // length of x
   x = malloc(nx*sizeof(double));
   for (i = 0; i < nx; i++)
      x[i] = rand() / (double) RAND_MAX;
   double startime,endtime;
   startime = omp_get_wtime();
   // parallel
   burst(x,nx,k,&startmax,&endmax,&maxval);
   // back to single thread
   endtime = omp_get_wtime();
   printf("elapsed time:  %f\n",endtime-startime);
   printf("%d %d %f\n",startmax,endmax,maxval);
   if (nx < 25) {
      for (i = 0; i < nx; i++) printf("%f ",x[i]);
      printf("\n");
   }
}
```

5.2.2 Compiling and Running

One does need to specify to the compiler that one is using OpenMP. On Linux, for instance, I compiled the code via the command

```
% gcc -g -o burst Burst.c -fopenmp -lgomp
```

Here I am directing the compiler **gcc** to process my C source file **Burst.c**, producing as output the executable file **burst**. I specify **-fopenmp** to warn the compiler that I am using the OpenMP pragmas, and ask to link in the OpenMP runtime library, **gomp**. For debugging purposes, I include the flag **-g**.

Note too that there is a corresponding include-file line in the code, to include the OpenMP definitions:

```
#include <omp.h>
```

Here is a sample run k = 10 and n = 2500:

```
% burst 3 10
elapsed time:  0.000626
2 4 0.831062
0.840188 0.394383 0.783099 0.798440 0.911647 0.197551
  0.335223 0.768230 0.277775 0.553970
```

5.2.3 Analysis

Now, take a look at **burst()**:

```
void burst(double *x, int nx, int k,
   int *startmax, int *endmax, double *maxval)
   {
   int nth;   // number of threads
   #pragma omp parallel
   {  int perstart, // START OF PARALLEL BLOCK
          perlen,  // period length
   ...
   ...
   ...
              *startmax = mystartmax;
              *endmax = myendmax;
         }
      }
   }  // END OF PARALLEL BLOCK
```

This is really the crux of OpenMP. Note the *pragma*:

```
#pragma omp parallel
```

This instruction to the compiler unleashes a team of threads. Each of the threads will execute the block that follows,[2] with certain rules governing the local variables:

Consider the variable **nth**.[3] It is local to **burst()**, but significantly it is *outside* the block executed by the threads. This means, in effect, that **nth** acts globally from the point of view of the threads, with this variable being shared by all the threads. If one thread changes the value of this variable, the other threads see the new value if they read **nth**.

By contrast, **perstart** is declared *inside* the threads block. This means that each thread will have its own **perstart**, acting completely independently of the others; this variable is *not* shared.

Shared-memory programming, by definition, needs shared variables. In threads programming, all the global variables are shared, but the above

[2]A *block* in C/C++ consists of code contained between left and right braces, { and }. Here, we've highlighted them with START... and END... comments.

[3]Actually, this variable is not used. However, I usually include it for debugging purposes.

scope rules give the programmer the ability to designate some nonglobals as shared as well. (OpenMP also has other options for this, which will not be covered here.)

Let's look at the next pragma:

```
#pragma omp single
{
    nth = omp_get_num_threads();
}
```

The **single** pragma directs that one thread (whichever reaches this line first) will execute the next block, while the other threads wait. In this case, we are just setting **nth**, the number of threads, and since the variable is shared, only one thread need set it.

As mentioned, the other threads will wait for the one executing that **single** block. In other words, there is an implied barrier right after the block. In fact, OpenMP inserts invisible barriers after all **parallel**, **for** and **sections** pragma blocks. In some settings, the programmer knows that such a barrier is unnecessary, and can use the **nowait** clause to instruct OpenMP to not insert a barrier after the block:

```
#pragma omp for nowait
```

Of course, programmers may need to insert their own barriers at various places in their code. The OpenMP **barrier** pragma is available for this.

As usual, we need each thread to know its own ID number:

```
me = omp_get_thread_num();
```

Note again that **me** was declared *inside* the **parallel** pragma block, so that each thread will have a different, independent version of this variable—which of course is exactly what we need.

Unlike most of our earlier examples, the code here does not break our data into chunks. Instead, the workload is partitioned in a different way to the threads. Here is how. Look at the nested loop,

```
for (perstart = 0; perstart <= nx-k; perstart++) {
   for (perlen = k; perlen <= nx - perstart; perlen++) {
```

The outer loop iterates over all possible starting points for a burst period, while the inner loop iterates over all possible lengths for the period. One natural way to divide up the work among the threads is to parallelize the outer loop. The **for** pragma does exactly that:

```
#pragma omp for
for (perstart = 0; perstart <= nx-k; perstart++) {
```

This pragma says that the following **for** loop will have its iterations divided among the threads. Each thread will work on a separate set of iterations, thus accomplishing the work of the loop in parallel. (Clearly, a requirement is that the iterations must be independent of each other.) One thread will work on some values of **perstart**, a second thread will work on some other values, and so on.

Note that we won't know ahead of time which threads will handle which loop iterations. We'll have more on this below, but the point is that there will be some partitioning done by the OpenMP code, thus parallelizing the computation. Of course, a **for** pragma is meaningless if it is not inside a **parallel** block, as there would be no threads to assign the iterations to.

The way we've set things up here, the inner loop,

```
for (perlen = k; perlen <= nx - perstart; perlen++) {
```

does not have its work partitioned among threads. For any given value of **perstart**, all values of **perlen** will be handled by the same thread.

So, each thread will keep track of its own record values, i.e., the location and value of the maximal burst it has found so far. In the end, each thread will need to update the overall record values, in this code:

```
if (mymaxval > *maxval) {
   *maxval = mymaxval;
   *startmax = mystartmax;
   *endmax = myendmax;
}
```

This is a critical section, and the code must be executed atomically. If we were programming directly with a threads interface library, we would need to declare a lock variable and initialize the lock at the beginning of the function **burst()**, and then have code locking and unlocking the lock immediately before and after the critical section. By contrast, a programmer's life is much easier with OpenMP: One simply inserts an OpenMP **critical** pragma:

```
#pragma omp critical
{
   if (mymaxval > *maxval) {
      *maxval = mymaxval;
      *startmax = mystartmax;
      *endmax = myendmax;
   }
}
```

5.2.4 A Cautionary Note About Thread Scheduling

In Section 5.2.3, it was stated, concerning the code

```
#pragma omp parallel
```

"This instruction to the compiler unleashes a team of threads." This sounds innocuous, but what does it really mean? All it says is that the threads will be created. They will now appear in the operating system's process table in *ready* state (Section 2.6)—in other words, they are not necessarily actually running yet.

The order in which the threads do start running is random, unpredictable. It's quite important to keep this in mind, as subtle and hard-to-fix bugs may occur if it is ignored.

5.2.5 Setting the Number of Threads

One can set the number of threads either before or during execution, For the former, one sets the **OMP_NUM_THREADS** environment variable, e.g.

```
export OMP_NUM_THREADS=8
```

to specify 8 threads in the **bash** shell on Unix-family systems. To do this programmatically, use **omp_set_num_threads()**.

Technically, these only specify an upper bound on the number of threads used. The OpenMP runtime system may choose to override the specified value with a smaller number. You can disable this by

```
omp_set_dynamic(0)
```

# threads	time
2	18.543343
4	11.042197
8	6.170748
16	3.183520

Table 5.1: Timings for the maximal-burst example

5.2.6 Timings

Timings on simulated data, with n = 50000 and k = 100, on the 16-core machine described in this book's Preface, are shown in Table 5.1. The pattern was fairly linear, with each doubling in the number of threads producing an approximate halving of run time.

5.3 OpenMP Loop Scheduling Options

You may have noticed that we have a potential load balance problem in the above maximal-burst example. Iterations that have a larger value of **perstart** do less work. In fact, the pattern here is very similar to that of our mutual outlinks example, in which we first mentioned the load balance issue (Section 1.4.5.2). Thus the manner in which iterations are assigned to threads may make a big difference in program speed.

So far, we haven't discussed the details of how the various iterations in a loop are assigned to the various threads. Back in Section 3.1, we discussed general strategies for doing this, and OpenMP offers the programmer several options along those lines.

5.3.1 OpenMP Scheduling Options

The type of scheduling is specified via the **schedule** clause in a **for** pragma, e.g.

```
#pragma omp for schedule(static)
```

and

```
#pragma omp for schedule(dynamic,50)
```

The keywords **static** and **dynamic** correspond to the scheduling strategies presented in Section 3.1, with the optional second argument being chunk size as discussed in that section. The static version assigns chunks before the loop is executed, parceled out in Round Robin manner.

The third scheduling option is **guided**. It uses a large chunk size in early iterations, but tapers down the chunk size as the execution of the loop progresses. This strategy, also discussed in Section 3.1, is designed to minimize overhead in the early rounds, but minimize load imbalance later on. Details are implementation-dependent.

Instead of hardcoding the options as above, one can allow the choices to be made a run time, either via the function **omp_set_schedule()** or by setting the environment variable **OMP_SCHEDULE**.

Continuing the timing experiments from Section 5.2.6, with k = 10 and n = 75000, produced the results in Table 5.2.

Not much pattern emerges. There did seem to be a penalty for using too large a chunk size with 4 threads, probably reflecting load imbalance.

And most importantly, the default settings seem to work well. Unfortunately, they are implementation-dependent, but things at least worked well on this platform (GCC version 4.6.3 on Ubuntu).

As a rule of thumb, fine-tuning schedule settings should make a difference only in very special applications. For example, if one has a small number of threads, a small number of iterations and the iteration times are large and widely-varying (in unpredictable ways), one might try a dynamic schedule with a chunk size of 1.

5.3.2 Scheduling through Work Stealing

We note some OpenMP-like systems that do internal *work stealing*, such as Threading Building Blocks (TBB, Section 5.11) and Cilk++. Their internal algorithms for partitioning work to threads are aimed at providing better load balance. The algorithms do runtime checks to see whether one thread has become idle while another thread has a queue of work to do. In such a case, work is transferred from the overburdened thread to the idle one—all without the programmer having to go to any effort.

# theads	sched, chunk	time
4	default	22.773100
4	static, 1	22.932213
4	static, 50	22.887986
4	static, 500	25.730284
4	dynamic, 1	22.841720
4	dynamic, 50	22.774348
4	dynamic, 500	23.669525
4	guided	22.767232
16	default	7.081358
16	static, 1	7.046007
16	static, 50	7.059683
16	static, 500	7.010607
16	dynamic, 1	7.060027
16	dynamic, 50	7.020815
16	dynamic, 500	7.010607
16	guided	7.194322

Table 5.2: Timings, for various scheduling options

Again, for most looping applications this won't be necessary. But for complicated algorithms with dynamic work queues, work stealing may produce a performance boost.

5.4 Example: Transforming an Adjacency Matrix

Let's see how the example in Section 4.8 can be implemented in OpenMP.

(It is recommended that the reader review the R version of this algorithm before continuing. The pattern used below is similar, but a bit harder to follow in C, which is a lower-level language than R.)

5.4.1 The Code

```
// AdjMatXform.c

// takes a graph adjacency matrix for a directed
// graph, and converts it to a 2-column matrix of
// pairs (i,j), meaning an edge from vertex i to
// vertex j; the output matrix must be in
// lexicographical order

#include <omp.h>
#include <stdlib.h>
#include <stdio.h>

// transgraph() does the work
// arguments:

//     adjm:  the adjacency matrix (NOT assumed
//            symmetric), 1 for edge, 0 otherwise;
//            matrix is overwritten by the function
//     n:  number of rows and columns of adjm
//     nout:  output, number of rows in returned matrix
// returned value: pointer to the converted matrix

// finds chunk among 0,...,n-1 to assign to thread
// number me among nth threads
void findmyrange(int n,int nth,int me,int *myrange)
```

```
{  int chunksize = n / nth;
   myrange[0] = me * chunksize;
   if (me < nth-1)
      myrange[1] = (me+1) * chunksize - 1;
   else myrange[1] = n - 1;
}

int *transgraph(int *adjm, int n, int *nout)
{
   int *outm,  // to become the output matrix
       *num1s,  // i-th element will be number of 1s
                // in row i of adjm
       *cumul1s;  // cumulative sums in num1s
   #pragma omp parallel
   {  int i,j,m;
      int me = omp_get_thread_num(),
          nth = omp_get_num_threads();
      int myrows[2];
      int tot1s;
      int outrow,num1si;
      #pragma omp single
      {
         num1s = malloc(n*sizeof(int));
         cumul1s = malloc((n+1)*sizeof(int));
      }
      // determine the rows in adjm to be handled
      // by this thread
      findmyrange(n,nth,me,myrows);
      // now go through each row of adjm assigned
      // to this thread, recording the locations
      // (column numbers) of the 1s; to save on
      // malloc() ops, reuse adjm, writing locations
      // found in row i back into that row
      for (i = myrows[0]; i <= myrows[1]; i++) {
         // number of 1s found in this row
         tot1s = 0;
         for (j = 0; j < n; j++)
            if (adjm[n*i+j] == 1) {
               adjm[n*i+(tot1s++)] = j;
            }
         num1s[i] = tot1s;
      }
      // one thread will use num1s, set by all
```

```
      // threads so make sure they're all done
      #pragma omp barrier
      #pragma omp single
      {
         // cumul1s[i] tot 1s before row i of adjm
         cumul1s[0] = 0;
         // now calculate where the output of
         // each row in adjm should start in outm
         for (m = 1; m <= n; m++) {
            cumul1s[m] = cumul1s[m-1] + num1s[m-1];
         }
         *nout = cumul1s[n];
         outm = malloc(2*(*nout) * sizeof(int));
      }
      // implied barrier after "single" pragma
      // now fill in this thread's portion of the
      // output matrix
      for (i = myrows[0]; i <= myrows[1]; i++) {
         // current row within outm
         outrow = cumul1s[i];
         num1si = num1s[i];
         for (j = 0; j < num1si; j++) {
            outm[2*(outrow+j)] = i;
            outm[2*(outrow+j)+1] = adjm[n*i+j];
         }
      }
   }
   // implied barrier after "parallel" pragma
   return outm;
}
```

5.4.2 Analysis of the Code

Before we begin, note that parallel C/C++ code involving matrices typically is written in one dimension, as follows:

Consider a 3x8 array **x**. Since row-major order (recall Section 2.3) is used in C/C++, the array is stored internally in 24 consecutive words of memory, in row-by-row order. Keep in mind that C/C++ indices start at 0, not 1 as in R. The element in the second row and fifth column of the array is then **x[1,4]**, and it would be in the $8 + 4 = 12^{th}$ word in internal storage. In general, **x[i,j]** is stored in word

```
8 * i + j
```

of the array.

In writing generally-applicable code, we typically don't know at compile time how many columns (8 in the little example above) our matrix has. So it is typical to recognize the linear nature of the internal storage, and use it in our C code explicitly, e.g.

```
if (adjm[n*i+j] == 1) {
   adjm[n*i+(tot1s++)] = j;
```

The memory allocation issue has popped up again, as it did in the **Rdsm** implementation. Recall that in the latter, we allocated memory for an output of size equal to that of the worst possible case. In this case, we have chosen to allocate memory during the midst of execution, rather than allocating beforehand, with an array **num1s** that will serve the following purpose.

Note that if some row in the input matrix contains, say, five 1s, then this row will contribute five rows in the output. We calculate such information for each input row, placing this information in the array **num1s**:

```
for (i = myrows[0]; i <= myrows[1]; i++) {
   tot1s = 0;  // number of 1s found in this row
   for (j = 0; j < n; j++)
      if (adjm[n*i+j] == 1) {
         adjm[n*i+(tot1s++)] = j;
      }
   num1s[i] = tot1s;
}
```

Once that array is known, we find its cumulative values. These will inform each thread as to where that thread will write to the output matrix, and also will give us the knowledge of how large the output matrix will be. The latter information is used in the call to the C library memory allocation function **malloc()**:

```
#pragma omp barrier
#pragma omp single
{
   cumul1s[0] = 0;  // cumul1s[i] will be tot 1s before row i of adjm
   // now calculate where the output of each row in adjm
   // should start in outm
   for (m = 1; m <= n; m++) {
      cumul1s[m] = cumul1s[m-1] + num1s[m-1];
```

```
   }
   *nout = cumul1s[n];
   outm = malloc(2*(*nout) * sizeof(int));
}
```

Note again that memory allocation can be expensive, so in this particular implementation, we have decided to save allocation time (and space) by reusing **adjm** for scratch space. Thus the input matrix is written over, and would have to be saved before the call if it were still needed. Those intermediate results stored in the reused parts of **adjm**, which were the column numbers of the 1s that were found, are then used to fill out the output matrix:

```
// now fill in this thread's portion of the output matrix
for (i = myrows[0]; i <= myrows[1]; i++) {
   outrow = cumul1s[i];  // current row within outm
   num1si = num1s[i];
   for (j = 0; j < num1si; j++) {
      outm[2*(outrow+j)] = i;
      outm[2*(outrow+j)+1] = adjm[n*i+j];
   }
}
```

Note that implied and explicit barriers are used in this program. For instance, consider the second **single** pragma:

```
...
      }
   num1s[i] = tot1s;
}
#pragma omp barrier
#pragma omp single
{
   cumul1s[0] = 0;  // cumul1s[i] will be tot 1s before row i of adjm
   // now calculate where the output of each row in adjm
   // should start in outm
   for (m = 1; m <= n; m++) {
      cumul1s[m] = cumul1s[m-1] + num1s[m-1];
   }
   *nout = cumul1s[n];
   outm = malloc(2*(*nout) * sizeof(int));
}
for (i = myrows[0]; i <= myrows[1]; i++) {
   outrow = cumul1s[i];
...
```

The **num1s** array is used within the **single** pragma, but computed just before it. We thus needed to insert a barrier before the pragma, to make sure **nums1** is ready.

Similarly, the **single** pragma computes **cumul1s**, which is used by all threads after the pragma. Thus a barrier is needed right after the pragma, but OpenMP inserts an implicit barrier there for us, so we don't have an explicit one.

Note the OpenMP construct used in the program in Section 5.2.1 that is now missing in our current example—the **omp for** pragma. Here we specifically assign certain threads to certain (contiguous) rows of our adjacency matrix, something the **omp for** pragma would not give us.

5.5 Example: Adjacency Matrix, R-Callable Code

A typical application might involve an analyst writing most of his code in R, for convenience, but write the parallel part of the code in C/C++, to maximize speed. The most common interfaces for this are the R functions **.C()**, **.Call()** and **Rcpp**. We'll illustrate that notion here, modifying our earlier code for transforming an adjacency matrix, in Section 5.4.1.

5.5.1 The Code, for .C()

Code suitable for the **.C()** interface follows below.

```
// AdjMatXformForR.c

#include <R.h>
#include <omp.h>
#include <stdlib.h>

// transgraph() does this work
// arguments:
//    adjm:  the adjacency matrix (NOT assumed
//           symmetric), 1 for edge, 0 otherwise;
//           note: matrix is overwritten
//    np:   pointer to number of rows and
//          columns of adjm
//    nout:  output, number of rows in
//           returned matrix
//    outm:  the converted matrix
```

```
void findmyrange(int n,int nth,int me,int *myrange)
{  int chunksize = n / nth;
   myrange[0] = me * chunksize;
   if (me < nth-1)
      myrange[1] = (me+1) * chunksize - 1;
   else myrange[1] = n - 1;
}

void transgraph(int *adjm, int *np, int *nout,
   int *outm)
{
   int *num1s,   // i-th element will be the number
                 // of 1s in row i of adjm
       *cumul1s, // cumulative sums in num1s
       n = *np;
   #pragma omp parallel
   {  int i,j,m;
      int me = omp_get_thread_num(),
          nth = omp_get_num_threads();
      int myrows[2];
      int tot1s;
      int outrow,num1si;
      #pragma omp single
      {
         num1s = malloc(n*sizeof(int));
         cumul1s = malloc((n+1)*sizeof(int));
      }
      findmyrange(n,nth,me,myrows);
      for (i = myrows[0]; i <= myrows[1]; i++) {
         tot1s = 0;  // number of 1s found in this row
         for (j = 0; j < n; j++)
            if (adjm[n*j+i] == 1) {
               adjm[n*(tot1s++)+i] = j;
            }
         num1s[i] = tot1s;
      }
      #pragma omp barrier
      #pragma omp single
      {
         // cumul1s[i] will be tot 1s before
         // row i of adjm
         cumul1s[0] = 0;
         // now calculate where the output of each
```

```
            // row in adjm should start in outm
            for (m = 1; m <= n; m++) {
               cumul1s[m] = cumul1s[m-1] + num1s[m-1];
            }
            *nout = cumul1s[n];
         }
         int n2 = n * n;
         for (i = myrows[0]; i <= myrows[1]; i++) {
            // current row within outm
            outrow = cumul1s[i];
            num1si = num1s[i];
            for (j = 0; j < num1si; j++) {
               outm[outrow+j] = i + 1;
               outm[outrow+j+n2] = adjm[n*j+i] + 1;
            }
         }
      }
}
```

We could have a **main()** function here, but instead will be calling the code from R, as will be seen shortly.

5.5.2 Compiling and Running

In writing a C file **y.c** containing a function **f()** that we'll call from R, one can compile using R from a shell command line:

```
R CMD SHLIB y.c
```

This produces a runtime-loadable library file. In Linux or Mac systems, for instance, the file **y.so** would be created (possibly, with the corresponding file for Windows being **y.dll**. We then load it from R:

```
> dyn.load("y.so")
```

after which can call **f()** from R in some manner, such as **.C()** or **.Call()**. We've written the code above to be compatible with the simpler interface, **.C()**, which takes the form

```
> .C("f",our arguments here)
```

A more complex but more powerful call form, **.Call()** is also available, to be discussed below.

The file **y.c** must include the R header file:

```
#include <R.h>
```

Generally the good thing about compiling via R CMD SHLIB is that we don't have to worry where the header file is, or worry about the library files. But things are a bit more complicated if one's code uses OpenMP, in which case we must so inform the compiler. We can do this by setting the proper environment variable. For C code and the **bash** shell, for instance, we would issue the shell command

```
% export SHLIB_OPENMP_CFLAGS = -fopenmp
```

Here is a sample run, again in the R interactive shell, with the C file being **AdjMatXformForR.c**:

```
n <- 5
dyn.load("AdjMatXformForR.so")
a <- matrix(sample(0:1,n^2,replace=TRUE),ncol=n)
out <-.C("transgraph",as.integer(a),as.integer(n),
   integer(1), integer(2*n^2))
```

Compare this last line to the signature of **transgraph()**:

```
void transgraph(int *adjm, int *np, int *nout, int *outm)
```

Note the following:

- The return value must be of type **void**, and in fact return values are passed via the arguments, in this case **nout** (the number of rows in the output matrix) and **outm** (the output matrix itself).
- All arguments are pointers.
- Our R code must allocate space for the output arguments.

Concerning that last point, there is no longer reason to have our C code allocate memory for the output matrix, as it did in Section 5.4. Here we set up that matrix to have worst-case size before the call, as we did in the **Rdsm** version.

So, here is a test run:

```
> n <- 5
> dyn.load("AdjMatXformForR.so")
> a <- matrix(sample(0:1,n^2,replace=TRUE),ncol=n)
> out <-.C("transgraph",as.integer(a),as.integer(n),
+  integer(1),integer(2*n^2))
> out
[[1]]
 [1] 0 0 0 1 0 1 3 0 4 1 3 4 0 0 3 4 1 0 0 4 1 1 0 1 1

[[2]]
[1] 5

[[3]]
[1] 14

[[4]]
 [1] 1 1 1 1 2 2 2 3 4 4 5 5 5 5 0 0 0 0 0 0 0 0 0 0 0 1
 2 4 5 1 4 5 1 2 5 1 2 4
[39] 5 0 0 0 0 0 0 0 0 0 0 0
```

As you can see, the return value of **.C()** is an R list, with one element for each of the arguments to **transgraph()**, including the output arguments.

Note that by default, all input arguments are duplicated, so that any changes to them are visible only in the output list, not the original arguments. Here **out**[[**1**]] is different from the input matrix **a**:

```
> a
     [,1] [,2] [,3] [,4] [,5]
[1,]    1    1    0    1    1
[2,]    1    0    0    1    1
[3,]    1    0    0    0    0
[4,]    0    1    0    0    1
[5,]    1    1    0    1    1
```

Duplication of the data might impose some slowdown, and can be disabled, but this usage is discouraged by the R development team.

Our output matrix, **out**[[**4**]], is hard to read in its linear form. Let's display it as a matrix, keeping in mind that our other output variable, **out**[[**3**]], tells us how many (real) rows there are in our output matrix:

```
> (nout <- out[[3]])
[1] 14
```

```
> o4 <- out[[4]]
> om <- matrix(o4,ncol=2)
> om[1:nout,]
      [,1] [,2]
 [1,]    1    1
 [2,]    1    2
 [3,]    1    4
 [4,]    1    5
 [5,]    2    1
 [6,]    2    4
 [7,]    2    5
 [8,]    3    1
 [9,]    4    2
[10,]    4    5
[11,]    5    1
[12,]    5    2
[13,]    5    4
[14,]    5    5
```

5.5.3 Analysis

So, what has changed in this version? Most of the change is due to the differences between R and C.

Most importantly, the fact that R uses column-major storage for matrices while C uses row-major order (Section 2.3) means that much of our new code must "reverse" the old code. For example, the line

```
outm[2*(outrow+j)+1] = adjm[n*i+j];
```

in the original code now becomes

```
int n2 = n * n;
...
outm[outrow+j+n2] = adjm[n*j+i] + 1;
```

5.5.4 The Code, for Rcpp

The other major way to call C/C++ code from R is via the **.Call()** function. It is considered more advanced than **.C()**, but is much more complex. That complexity, though, can be largely hidden from the programmer through the use of the **Rcpp** package, and in fact the net result is that the **Rcpp** route is actually easier than using **.C()**.

Here is the **Rcpp** version of our previous code:

```
// AdjRcpp.cpp

#include <Rcpp.h>
#include <omp.h>

// the function transgraph() does the work
// arguments:
//    adjm:  the adjacency matrix (NOT assumed
//           symmetric), 1 for edge, 0 otherwise;
//            note: matrix is overwritten
//           by the function
//  return value: the converted matrix

// finds the chunk of rows this thread will process
void findmyrange(int n,int nth,int me,int *myrange)
{  int chunksize = n / nth;
   myrange[0] = me * chunksize;
   if (me < nth-1)
      myrange[1] = (me+1) * chunksize - 1;
   else myrange[1] = n - 1;
}

RcppExport SEXP transgraph(SEXP adjm)
{
   int *num1s,  // i-th element will be the number
                // of 1s in row i of adjm
       *cumul1s,  // cumulative sums in num1s
       n;
   Rcpp::NumericMatrix xadjm(adjm);
   n = xadjm.nrow();
   int n2 = n*n;
   Rcpp::NumericMatrix outm(n2,2);

   #pragma omp parallel
   {  int i,j,m;
      int me = omp_get_thread_num(),
          nth = omp_get_num_threads();
      int myrows[2];
```

```
      int tot1s;
      int outrow,num1si;
      #pragma omp single
      {
         num1s = (int *) malloc(n*sizeof(int));
         cumul1s = (int *) malloc((n+1)*sizeof(int));
      }
      findmyrange(n,nth,me,myrows);
      for (i = myrows[0]; i <= myrows[1]; i++) {
         // number of 1s found in this row
         tot1s = 0;
         for (j = 0; j < n; j++)
            if (xadjm(i,j) == 1) {
               xadjm(i,(tot1s++)) = j;
            }
         num1s[i] = tot1s;
      }
      #pragma omp barrier
      #pragma omp single
      {
         // cumul1s[i] will be tot 1s before row
         // i of xadjm
         cumul1s[0] = 0;
         // now calculate where the output of each
         // row in adjm should start in outm
         for (m = 1; m <= n; m++) {
            cumul1s[m] = cumul1s[m-1] + num1s[m-1];
         }
      }
      for (i = myrows[0]; i <= myrows[1]; i++) {
         // current row within outm
         outrow = cumul1s[i];
         num1si = num1s[i];
         for (j = 0; j < num1si; j++) {
            outm(outrow+j,0) = i + 1;
            outm(outrow+j,1) = xadjm(i,j) + 1;
         }
      }
   }

   Rcpp::NumericMatrix outmshort =
      outm(Rcpp::Range(0,cumul1s[n]-1),
         Rcpp::Range(0,1));
```

```
   return outmshort;
}
```

5.5.5 Compiling and Running

We will still run **R CMD SHLIB** to compile, but we have more libraries to specify in this case. In the **bash** shell, we can run

```
export R_LIBS_USER=/home/nm/R
export PKG_LIBS="-lgomp"
export PKG_CXXFLAGS="-fopenmp -I/home/nm/R/Rcpp/include"
```

That first command lets R know where our R packages are, in this case the **Rcpp** package. The second states we need to link in the **gomp** library, which is for OpenMP, and the third both warns the compiler to watch for OpenMP pragmas and to include the **Rcpp** header files.

Note that that last **export** assumes our source code is in C++, as indicated below by a **.cpp** suffix to the file name. Since C is a subset of C++, our code can be pure C but we are presenting it as C++.

We then run

```
R CMD SHLIB AdjRcpp.cpp
```

This produces a **.so** file or equivalent as before. Here is a sample run:

```
> library(Rcpp)  # don't forget to do this FIRST
> dyn.load("AdjRcpp.so")
> m <- matrix(sample(0:1,16,replace=TRUE),ncol=4)
> m
     [,1] [,2] [,3] [,4]
[1,]    1    1    1    0
[2,]    1    1    0    1
[3,]    1    1    0    0
[4,]    1    0    0    1
> .Call("transgraph",m)
     [,1] [,2]
[1,]    1    1
[2,]    1    2
```

```
 [3,]    1    3
 [4,]    2    1
 [5,]    2    2
 [6,]    2    4
 [7,]    3    1
 [8,]    3    2
 [9,]    4    1
[10,]    4    4
```

Sure enough, we do use **.Call()** instead of **.C()**. And note that we have only one argument here, **m**, rather than five as before, and that the result is actually in the return value, rather than being in one of the arguments. In other words, even though **.Call()** is more complex than **.C()**, use of **Rcpp** makes everything much simpler than under **.C()**. In addition, **Rcpp** allows us to write our C/C++ code as if column-major order were used, consistent with R. No wonder **Rcpp** has become so popular!

5.5.6 Code Analysis

The heart of using **.Call()**, including via **Rcpp**, is the concept of the SEXP ("S-expression," alluding to R's roots in the s language). In R internals, a SEXP is a pointer to a C struct containing the given R object and information about the object. For instance, the internal storage for an R matrix will consist of a struct that holds the elements of the matrix and and its numbers of rows and columns. It is this encapsulation of data and metadata into a struct that enabled us to have only a single argument in the new version of **transgraph()**:

```
RcppExport SEXP transgraph(SEXP adjm)
```

The term **RcppExport** will be explained shortly. But first, note that both the input argument, **adjm**, and the return value are of type SEXP. In other words, the input is an R object and the output is an R object. In our run example above,

```
> .Call("transgraph",m)
```

the input was the R matrix **m**, and the output was another R matrix.

The machinery in **.Call()** here is set up for C, and C++ users (including us in the above example) need a line like

```
extern "C" transgraph;
```

in the C++ code. The **RcppExport** term is a convenience for the programmer, and is actually

```
#define RcppExport extern "C"
```

Now, let's see what other changes have been made. Consider these lines:

```
Rcpp::NumericMatrix xadjm(adjm);
n = xadjm.nrow();
int n2 = n*n;
Rcpp::NumericMatrix outm(n2,2);
```

Rcpp has its own vector and matrix types, serving as a bridge between those types in R and corresponding arrays in C/C++. The first line above creates an **Rcpp** matrix **xadjm** from our original R matrix **adjm**. (Actually, no new memory space is allocated; here **xadjm** is simply a pointer to the data portion of the struct where **adjm** is stored.) The encapsulation mentioned earlier is reflected in the fact that **Rcpp** matrices have the built-in method **nrow()**, which we use here. Then we create a new $n^2 \times 2$ **Rcpp** matrix, **outm**, which will serve as our output matrix. As before, we are allowing for the worst case, in which the input matrix consists of all 1s.

Rcpp really shines for matrix code. Recall the discussion at the beginning of Section 5.4.2. In our earlier versions of this adjacency matrix code, both in the standalone C and R-callable versions, we were forced to use one-dimensional subscripting in spite of working with two-dimensional arrays, e.g.

```
if (adjm[n*i+j] == 1) {
```

This was due to the fact that ordinary two-dimensional arrays in C/C++ must have their numbers of columns declared at compile time, whereas in this application such information is unknown until run time. This is not a problem with object-oriented structures, such as those in the C++ Standard Template Library (STL) and **Rcpp**.

So now with **Rcpp** we can use genuine two-dimensional indexing, albeit with parentheses instead of brackets:[4]

```
if (xadjm(i,j) == 1) {
```

Note, though, that **Rcpp** subscripts follow C/C++ style, starting at 0 rather than 1 for R. The "+1" in

[4]It is still possible to do one-dimensional indexing, using brackets, but recall that **Rcpp** uses column-major order for compatibility with R.

```
outm(outrow+j,1) = xadjm(i,j) + 1;
```

in which we were inserting a certain column number from the adjacency matrix, was needed to resolve this discrepancy.

Most of the remaining code is unchanged, except for the return value:

```
Rcpp::NumericMatrix outmshort =
   outm(Rcpp::Range(0,cumul1s[n]-1),Rcpp::Range(0,1));
return outmshort;
```

As before, we allocated space for **outm** to allow for the worst case, in which n^2 rows were needed. Typically, there are far fewer than n^2 1s in the matrix **adjm**, so the last rows in **outm** are filled with 0s. Here we copy the nonzero rows into a new **Rcpp** matrix **outmshort**, and then return that.

All in all, **Rcpp** made our code simpler and easier to write: We have fewer arguments, arguments are in explicit R object form, we don't need to deal with row-major vs. column-major order, and our results come back in exactly the desired R object, rather than as one component of a returned R list.

5.5.7 Advanced Rcpp

Rcpp is at this writing becoming increasingly more versatile, offering several ways to set up code, other than the very basic approach presented here.

One advanced feature (many would consider it basic, not advanced) is Rcpp *attributes*, which enables simpler code, though with an extra build step. For instance, the argument **adjm** in **transgraph()** could be declared as of type **Rcpp::NumericMatrix** rather than SEXP. This would be clearer, and would save us the trouble of creating an extra variable, **xadjm**.

Another example is Rcpp *syntactic sugar*, which magically allows you to add some R-style syntax to C++, very nice!

5.6 Speedup in C

So, let's check whether running in C can indeed do much better than R in a parallel context, as discussed back in Section 1.1.

```
> n <- 10000
```

cores	language	time
1	R	27.516
4	R (Rdsm)	7.783
4	C (OpenMP)	3.193

Table 5.3: Timing comparisons

```
> a <- matrix(sample(0:1,n^2,replace=TRUE),ncol=n)
> system.time(out <-.C("transgraph",as.integer(a),
+    as.integer(n),integer(1),integer(2*n^2)))
   user  system elapsed
  5.692   0.852   3.193
```

Gathering our old timings, the various methods are compared in Table 5.3.

5.7 Run Time vs. Development Time

Inspecting Table 5.3, we see that going from serial R to parallel R cut down run time by about 72%, while the corresponding figure for OpenMP was 88%. To be sure, the OpenMP version was actually more than twice as fast as the parallel R one. But *relative to the serial R code*, the move to C yielded only a modest improvement over parallel R.

We thus see here a concrete illustration of the Principle of Pretty Good Parallelism introduced in Section 1.1: Running in C can indeed pay off, if we are willing to devote the development time, but that payoff may not be worth the effort.

5.8 Further Cache/Virtual Memory Issues

It has been mentioned several times in this book that cache coherency transactions (and virtual memory paging) can really compromise performance. Coupling that with the point, made in Section 2.3.4, that different designs of the same code can have quite different memory access patterns and thus quite different cache performance, we see that we must be mindful of such issues when we write shared-memory code. And remember, the problem is especially acute in multicore settings, due to cache coherency issues (Section 2.5.1.1).

To make this idea concrete, we'll look at two OpenMP programs to do in-place matrix transpose. Here's the first:

```
// CacheRow.c

#include <omp.h>
#include <stdlib.h>
#include <stdio.h>

// translate from 2-D to 1-D indices
int onedim(int n,int i,int j) {  return n * i + j;  }

void transp(int *m, int n)
{
   #pragma omp parallel
   {  int i,j,tmp;
      // walk through all the above-diagonal elements,
      // swapping them with their below-diagonal mates
      #pragma omp for
      for (i = 0; i < n; i++) {
         for (j = i+1; j < n; j++) {
            tmp = m[onedim(n,i,j)];
            m[onedim(n,i,j)] = m[onedim(n,j,i)];
            m[onedim(n,j,i)] = tmp;
         }
      }
   }
}

int *m;

int main(int argc, char **argv)
{  int i,j;
   int n = atoi(argv[1]);
   m = malloc(n*n*sizeof(int));
   for (i = 0; i < n; i++)
      for (j = 0; j < n; j++)
         m[n*i+j] = rand() % 24;
   if (n <= 10) {
      for (i = 0; i < n; i++) {
```

```
         for (j = 0; j < n; j++) printf("%d ",m[n*i+j]);
         printf("\n");
      }
   }
   double startime,endtime;
   startime = omp_get_wtime();
   transp(m,n);
   endtime = omp_get_wtime();
   printf("elapsed time:  %f\n",endtime-startime);
   if (n <= 10) {
      for (i = 0; i < n; i++) {
         for (j = 0; j < n; j++)
            printf("%d ",m[n*i+j]);
         printf("\n");
      }
   }
}
```

The code is fairly straightforward. It goes through the matrix row-by-row, exchanging the above-diagonal elements of each row with their corresponding below-diagonal elements.

Recall once again that C stores matrices in row-major order. So, as the above code traverses the matrix, it is staying in the same cache block for a sustained amount of time, i.e., the cache performance is fairly good. We say only "fairly" here, as the below-diagonal elements are being traversed column-by-column, thus not auguring well for cache performance. Nevertheless, it would seem that this code will do better than the second version:

```
// CacheWave.c

#include <omp.h>
#include <stdlib.h>
#include <stdio.h>

// translate from 2-D to 1-D indices
int onedim(int n,int i,int j) {  return n * i + j;  }

void trade(int *m,int n,int i,int j) {
   int tmp;
   tmp = m[onedim(n,i,j)];
   m[onedim(n,i,j)] = m[onedim(n,j,i)];
   m[onedim(n,j,i)] = tmp;
}
```

```
void transp(int *m, int n)
{  int n1 = n - 1;
   int n2 = 2 * n - 3;
   #pragma omp parallel
   {  int w,j;
      int row, col;
      #pragma omp for
      // w is wavefront number, indexed across
      // top row, bottom row
      // we move from northeast to southwest
      // within diagonals
      for (w = 1; w <= n2; w++) {
         if (w < n) {
            row = 0;
            col = w;
         } else {
            row = w - n1;
            col = n1;
         }
         for (j = 0; ; j++) {
            if (row > n1 || col < 0) break;
            if (row >= col) break;
            trade(m,n,row++,col--);
         }
      }
   }
}

int *m;

int main(int argc, char **argv)
{  int i,j;
   int n = atoi(argv[1]);
   m = malloc(n*n*sizeof(int));
   for (i = 0; i < n; i++)
      for (j = 0; j < n; j++)
         m[n*i+j] = rand() % 24;
   if (n <= 10) {
      for (i = 0; i < n; i++) {
         for (j = 0; j < n; j++)
            printf("%d ",m[n*i+j]);
         printf("\n");
```

```
      }
   }
   double startime,endtime;
   startime = omp_get_wtime();
   transp(m,n);
   endtime = omp_get_wtime();
   printf("elapsed time: %f\n",endtime-startime);
   if (n <= 10) {
      for (i = 0; i < n; i++) {
         for (j = 0; j < n; j++)
            printf("%d ",m[n*i+j]);
         printf("\n");
      }
   }
}
```

This version uses a *wavefront* approach. Here, instead of each iteration of the **for** loop processing a different row, each iteration now involves a different "northeast to southwest" anti-diagonal. For instance, consider the iteration **w = 3** in the outer **for** loop in **transp()**. It will process **m[0,3]**, **m[1,2]**, **m[2,1]** and **m[3,0]**.

Wavefront methods are widely used in matrix algorithms and can be very advantageous. Yet in this particular application, the memory usage pattern is more random from a caching point of view, and one suspects that the resulting poorer hit rate will adversely impact performance. In plain English: The second version should be slower.

Moreover, we would guess that, the more cores we use, the worse the speed discrepancy between the two versions of the program. Any cache miss on a write may cause cache operations at any of the other caches, and since we have a cache for each core, our troubles should intensify as system size grows.

This is confirmed in the timing experiments shown in Table 5.4. The matrix sizes were 25000x25000. We see right away that it does pay to be mindful of cache implications when one writes one's code. And sure enough, the more cores we use, the worse the ratio in run times between the two versions of code.

Programmers who spend time truly optimizing their code may go further, for instance worrying about *false sharing*. Suppose our code writes to a variable **x**, thus invalidating that particular cache block—which, recall, means the *entire* block. There may be a perfectly good copy of another variable **y**

# cores	rowwise	wavefront	ratio
4	9.119054	10.767355	0.8469168
8	4.874676	6.173957	0.7895546
16	2.586739	3.545786	0.7295249

Table 5.4: Timings: same application, different memory patterns

in the same block. Yet now an access to **y** will trigger an unnecessary and expensive cache coherency operation, since **y** is in a "bad" block.

One could avoid such a calamity by placing *padding* in between our declarations of **x** and **y**, say

```
int x,w[63],y;  // all assumed global
```

If our cache block size is 512 bytes, i.e., 64 8-byte integers, then **y** should be 512 bytes past **x** in memory, hence not in the same block.

5.9 Reduction Operations in OpenMP

A very common operation in parallel computation is *reduction*, in which a grand total or some other single number is computed from many. In addition to a sum, we might compute a minimum, a maximum and so on. Most programming languages and libraries for parallel computation include some kind of reduction construct. It has been mentioned earlier in this book, for instance, that R has a **Reduce()** function, and that most message-passing systems have something like that too. OpenMP offers this too, as we will see in this section.

5.9.1 Example: Mutual In-Links

Here we again revisit our mutual outlinks problem, from Section 1.4. In this case, we'll compute inbound links, partly for variety but also to make a point about caches in Section 5.9.2.

5.9.1.1 The Code

```
// MutInlinks.cpp

// mutual in-links prevalence computation

// input is a graph adjacency matrix, element (i,j)
// being 1 or 0, depending on whether there is an
// edge from vertex i to vertex j

// we find the total number of mutual in-links, over
// all possible pairs of vertices

#include <Rcpp.h>
#include <omp.h>

// count number of mutual inlinks among vertices
// j > i with vertex i, in the matrix xa of type
// Rcpp::NumericMatrix
int do_one_i(Rcpp::NumericMatrix xa,int i)
{
   int nr = xa.nrow();
   int nc = xa.ncol();
   int sum = 0,j,k;
   if (i >= nc - 1) return 0;
   for (j = i+1; j < nc; j++) {
      for (k = 0; k < nr; k++)
         sum += xa(k,i) * xa(k,j);
   }
   return sum;
}

RcppExport SEXP ompmutin(SEXP adj, SEXP nth)
{
   Rcpp::NumericMatrix xadj(adj);
   int nr = xadj.nrow();
   int nc = xadj.ncol();

   // set number of threads
   int nthreads = INTEGER(nth)[0];
   omp_set_num_threads(nthreads);

   // simplest approach
```

```
   int tot,i;
   #pragma omp parallel for reduction(+:tot)
   for (i = 0; i < nc; i++)
      tot += do_one_i(xadj,i);
   return Rcpp::wrap(tot);
}
```

5.9.1.2 Sample Run

```
> dyn.load("MutInlinks.so")
> m <- matrix(sample(0:1,25,replace=TRUE),ncol=5)
> m
     [,1] [,2] [,3] [,4] [,5]
[1,]    0    1    1    1    0
[2,]    1    0    0    1    1
[3,]    1    0    0    1    0
[4,]    1    1    1    0    0
> library(Rcpp)
> .Call("ompmutin",m,as.integer(2))
[1] 11
```

Note the need to write 2 in the call as **as.integer(2)**. This is not an **Rcpp** issue; instead, the problem is that R treats the constant 2 as having type **double**.

5.9.1.3 Analysis

The OpenMP reduction clause is seen here:

```
#pragma omp parallel for reduction(+:tot)
for (i = 0; i < nc; i++)
   tot += do_one_i(xadj,i);
```

In this pragma, note first that our **for** clause has in this case been accompanied by the **parallel** and **reduction** clauses. The former is there simply to save typing. The code

```
#pragma omp parallel
...
#pragma omp for
```

can be written more compactly as

```
#pragma omp parallel for
```

But the **reduction** clause is new to us here. In this case,

```
reduction(+:tot)
```

specifies that we will be computing a sum(+), storing it in the variable **tot**, and will do so in a "safe" manner, meaning the following. Note carefully that it may well occur that more than one thread is executing

```
tot += do_one_i(xadj, i);
```

at the same time, thus risking a race condition (Section 4.6.1). We need the above statement to be executed atomically.

We could make use of OpenMP's **critical** pragma (Section 5.2.1) to avoid this, but what's nice is that OpenMP does all that for us, behind the scenes, when we specify **reduction**. OpenMP would set up independent copies of **tot** for the various threads, and then add them atomically to the "real" **tot** when exiting the loop, but again, we need not worry about this.

5.9.2 Cache Issues

Any time a program is found to be slow, the first suspect is cache behavior. Often that suspicion is valid.

5.9.3 Rows vs. Columns

Recall that R uses column-major matrix storage, so that elements within the same column are stored contiguously. This implies that the mutual inlinks application should have better cache behavior than the mutual outlinks one, since in the former our code (at least in an easy, straightforward implementation) traverses columns rather than rows.

This also means that if we are interested in outlinks instead, it may pay to transpose the adjacency matrix before doing our analysis. That itself requires some time, but if we will be doing a lot of analysis on the matrix, in algorithms that also would normally traverse rows, this cost may be worthwhile.

5.9.4 Processor Affinity

Recall from Section 2.6 that when a thread starts a timeslice on a core, the cache at that core may not contain anything useful to that thread. The cache contents will then have to be built up, causing a lot of cache misses for a while, thus slowing things down.

It is thus desirable to be able to assign certain threads to certain cores, known as specifying *processor affinity.* If you use the **gcc** compiler, for instance, you can set the environment variable GOMP_CPU_AFFINITY for this. Or, from within an OpenMP program you can call **sched_setaffinity()**. Check the documentation for your system for details.

5.10 Debugging

Most debugging tools have the capability to follow specific threads. We'll use GDB here.

5.10.1 Threads Commands in GDB

First, as you run a program under GDB, the creation of new threads will be announced, e.g.

```
(gdb) r 100 2
Starting program: /debug/primes 100 2
[New Thread 16384 (LWP 28653)]
[New Thread 32769 (LWP 28676)]
[New Thread 16386 (LWP 28677)]
[New Thread 32771 (LWP 28678)]
```

You can do backtrace (**bt**) etc. as usual. Here are some threads-related commands:

- `info threads` (gives information on all current threads)
- `thread 3` (change to thread 3)
- `break 88 thread 3` (stop execution when thread 3 reaches source line 88)
- `break 88 thread 3 if x==y` (stop execution when thread 3 reaches source line 88 and the variables x and y are equal)

5.10.2 Using GDB on C/C++ Code Called from R

The basic idea is as follows:

- Start R itself under GDB.
- Place a breakpoint at the C/C++ function you wish to debug.
- Issue the **r** ("run") command to GDB, taking you to the R prompt.
- If you are using **Rcpp**, load that first.
- Call **dyn.load()** to bring in the C/C++ code.
- Run **.Call()**, resulting in GDB stopping in the desired C/C++ function.
- Then use GDB as usual.

Here is an example from the code in Section 5.2.1.

```
% R -d gdb
GNU gdb (GDB) 7.5.91.20130417-cvs-ubuntu
...
(gdb) b burst
Function "burst" not defined.
Make breakpoint pending on future shared library load? (y or [n]) y

Breakpoint 1 (burst) pending.
(gdb) r
Starting program: /usr/local/lib/R/bin/exec/R
[Thread debugging using libthread_db enabled]
Using host libthread_db library "/lib/i386-linux-gnu/libthread_db.so.1".

R version 3.0.1 (2013-05-16) -- "Good Sport"
...
> dyn.load("Burst.so")
> x <- sample(0:1,100,replace=TRUE)
> .Call("burst",x,as.integer(100),as.integer(10))

Breakpoint 1, burst (x=0x86c6b28, nx=141181456, k=141249936, startmax=0x0,
    endmax=0x86177d8, maxval=0x83b8b38) at Burst.c:27
27          {
(gdb) n
42                  nth = omp_get_num_threads();
...
```

5.11 Intel Thread Building Blocks (TBB)

At this writing (spring 2015), the closest competitor to OpenMP as a higher-level interface to threads is TBB, an open-source library developed by Intel. Here are some advantages and disadvantages:

- Due to work stealing (Section 5.3.2), and possibly better cache behavior, TBB code in some cases yields better performance.

- TBB does not require special compiler capability, in contrast say to OpenMP, which requires that the compiler understand OpenMP pragmas.

- TBB provides greater flexibility than OpenMP, though at the expense of considerably greater complexity.

Concerning this last bullet item, TBB requires one to make use of C++ *functors*, which are function objects taking the form of a struct or class. We will use these in the Thrust context in Chapter 7, but functors are just the beginning of the added complexity of TBB.

For instance, take reduction. A function **tbb::parallel_reduce()** is available in TBB, but it requires not only defining a "normal" functor, but also defining a second function with the struct or class, named **join()**.

Other than using TBB indirectly via Thrust, this book will not cover TBB. However, if you are a good C++ programmer, you may find TBB structure interesting and powerful. The principles of OpenMP covered in this chapter should provide a good starting point for you.

5.12 Lockfree Synchronization

Bear in mind that locks and barriers are "necessary evils." We do need them (or something equivalent) to ensure correct execution of our program, but they slow things down. For instance, we say that lock variables, or the critical sections they guard, *serialize* a program in the section they are used, i.e., they change its parallel character to serial; only one thread is allowed into the critical section at a time, so that execution is temporarily serial. And contention for locks can cause lots of cache coherency transactions, definitely putting a damper on performance. Thus one should always try to find clever ways to avoid locks and barriers if possible.

One way to do this is to take advantage of the hardware. Modern processors typically include a variety of hardware assists to make synchronization more efficient.

For example, Intel machines allow a machine instruction to be prefixed by a special byte called a **lock** prefix. It orders the hardware to lock up the system bus while the given instruction is executing—so that the execution is atomic. (The fact that this prefix, a hardware operation, is named **lock** should not be confused with lock variables in software.)

Under the critical section approach, code to atomically add 1 to **y** would look something like this:

```
lock the lock
add $1, y
unlock the lock
```

By contrast, we could do all this with a single machine instruction:

```
lock add $1, y
```

OpenMP includes an **atomic** pragma, which we'd use in the above example via this code:

```
#pragma omp atomic
y++;
```

This instructs the compiler to try to find a hardware construct like the **lock** prefix above to implement mutual exclusion, rather than taking the less efficient critical section route.

Also, the C++ Standard Template Library contains related constructs, such as the function **fetch_add()**, which again instructs the compiler to attempt to find an atomic hardware solution to the update-total example above. This idea has been advanced even further in C++11.

5.13 Further Reading

A number of books cover OpenMP in detail, such as *Multicore Application Programming*, by Darryl Gove, Addison-Wesley, 2011. For TBB, consult *Intel Threading Building Blocks*, by James Reinders, O'Reilly, 2007.

The book on **Rcpp**, *Seamless R and C++ Integration with Rcpp*, by Dirk Eddelbuettel, Springer, 2013, is useful but already quite outdated. See online sites for more recent enhancements to the package.

Chapter 6

The Shared-Memory Paradigm: GPUs

6.1 Overview

The video game market is so lucrative that the industry has developed ever-faster graphics cards, in order to handle ever-faster and ever-more visually detailed video games. These actually are parallel processing hardware devices, so around 2003 some people began to wonder if one might use them for parallel processing of nongraphics applications. Such programming was called *GPGPU*, general programming on graphics processing units, later shortened to simply GPU programming.

Originally this was cumbersome. One needed to figure out clever ways of mapping one's application to some kind of graphics problem, i.e., ways of disguising one's problem so that it appeared to be doing graphics computations. Though some high-level interfaces were developed to automate this transformation, effective coding required some understanding of graphics principles.

But current-generation GPUs separate out the graphics operations, and now consist of multiprocessor elements that run under the familiar shared-memory threads model. Granted, effective coding still requires an intimate knowledge of the hardware, but at least it's (more or less) familiar hardware, not requiring knowledge of graphics.

Moreover, unlike a multicore machine, with the ability to effectively run just a few threads at one time, e.g. four threads on a quad core machine, GPUs can run *hundreds or thousands* of threads well at once. There are various restrictions that come with this, but you can see that there is fantastic potential for speed here.

We will focus on NVIDIA's line of GPUs here. (For brevity, the presentation here will often refer to "GPUs" rather than "NVIDIA GPUs," but the latter is implied.) They are programmed in an extension of C/C++ called CUDA, for which various R interfaces have been developed. So as with OpenMP in Chapter 5, we again have an instance of the R+X notion introduced in Section 1.1.

To run the examples here, you'll need a CUDA-capable video card, and the CUDA development kit. To check whether your GPU is CUDA-capable, first determine what type you have (e.g. under Linux run **/sbin/lspci** or possibly **/usr/bin/lspci**), and then check the NVIDIA website or the CUDA Wikipedia entry.

The CUDA kit can be downloaded from the NVIDIA site. It is free of charge (though requires about a gigabyte of disk space).

6.2 Another Note on Code Complexity

Recall the discussion in Section 3.3, in which it was pointed out that efficient parallel programming often requires keen attention to detail. This is especially true for GPUs, which have a complex hardware structure, to be presented shortly. Thus, optimizing CUDA code is difficult.

Another issue is that NVIDIA is continuing to add more and more powerful features to that hardware. Thus we have a "moving target," making optimization even more of a challenge (though backward compatibility has been maintained).

On the other hand, it was noted at the start of Chapter 4 that in the parallel processing world, there is always a tradeoff between speed and programming effort. In many cases, one is happy to have just "good enough" speed, attained without having to expend herculean efforts in programming.

In light of this latter point, the CUDA code presented here is meant to have good speed while staying simple. But there is more:

An important approach to addressing the above issues is to make use of libraries. For many applications, very efficient CUDA libraries have been developed, such as the CUBLAS library for matrix operations. In addition, some R packages that include good CUDA code have also been developed. A big advantage of these, tying in with the "moving target" metaphor above, is that as NVIDIA hardware evolves, the libraries will typically be updated. This obviates the need for you to update your own CUDA code.

6.3 Goal of This Chapter

Thus, this chapter will first present examples directly written in CUDA, with relatively simple complexity but with reasonably good speed. These will serve the dual purposes of introducing CUDA

(a) for those who wish to write such code, and

(b) to illustrate how the hardware works, vital even for those who only use libraries.

6.4 Introduction to NVIDIA GPUs and CUDA

As noted, NVIDIA hardware and CUDA programming are threads- and shared-memory oriented. So far, so good, but there are wrinkles to this, so first brace yourself for a barrage of specialized terms: Execution at the GPU (termed the *device*) is *launched* at the CPU (the *host*), via a call to a programmer-written *kernel.* Note that the term *shared memory* in this context now refers to memory there on the graphics card, not the memory accessed by the CPU; this graphics card memory is called *global memory.* And to make matters even more confusing, there actually is something called *shared memory*, which you'll see really amounts to a cache. In the launch, the programmer also configures some special structures, *grid* and *blocks*, which determines how the threads are organized. All this will unfold in the next few pages.

The CUDA library includes routines to allocate space for data objects on the device, and to transfer data from the host to the device and vice versa. **Note carefully:** Such transfer can be a bottleneck in regard to efficiency of the code, and thus must be used carefully.

6.4.1 Example: Calculate Row Sums

Let's start with an easy one. Below is CUDA code that inputs a matrix and outputs an array consisting of the sums of the rows of the matrix.

```
// RowSums.cu; simple iilustration of CUDA

#include <stdio.h>
#include <stdlib.h>
#include <cuda.h>

// CUDA example:  finds row sums of an integer
//                matrix m

// find1elt() finds the row sum of one row of the
// nxn matrix m, storing the result in the
// corresponding position in the rowsum array rs;
// matrix is in 1-dimensional, row-major order

// this is the "kernel", which each thread on the
// GPU executes
__global__ void find1elt(int *m, int *rs, int n)
{
   // this thread will handle row # rownum
   int rownum = blockIdx.x;
   int sum = 0;
   for (int k = 0; k < n; k++)
      sum += m[rownum*n+k];
   rs[rownum] = sum;
}

// the remaining code is executed on the CPU
int main(int argc, char **argv)
{
    // number of matrix rows/cols
    int n = atoi(argv[1]);
    int *hm, // host matrix
        *dm, // device matrix
        *hrs, // host rowsums
        *drs; // device rowsums
    // size of matrix in bytes
    int msize = n * n * sizeof(int);
    // allocate space for host matrix
```

```
   hm = (int *) malloc(msize);
   // as a test, fill matrix with consec. integers
   int t = 0,i,j;
   for (i = 0; i < n; i++) {
      for (j = 0; j < n; j++) {
         hm[i*n+j] = t++;
      }
   }
   // allocate matrix space at device
   cudaMalloc((void **)&dm,msize);
   // copy host matrix to device matrix
   cudaMemcpy(dm,hm,msize,cudaMemcpyHostToDevice);
   // allocate host, device rowsum arrays
   int rssize = n * sizeof(int);
   hrs = (int *) malloc(rssize);
   cudaMalloc((void **)&drs,rssize);
   // set up threads structure parameters
   dim3 dimGrid(n,1);  // n blocks in the grid
   dim3 dimBlock(1,1,1);  // 1 thread per block
   // launch the kernel
   find1elt<<<dimGrid,dimBlock>>>(dm,drs,n);
   // wait until kernel finishes
   cudaThreadSynchronize();
   // copy row vector from device to host
   cudaMemcpy(hrs,drs,rssize,cudaMemcpyDeviceToHost);
   // check results
   if (n < 10)
      for(int i=0; i<n; i++) printf("%d\n",hrs[i]);
   // clean up, very important
   free(hm);
   cudaFree(dm);
   free(hrs);
   cudaFree(drs);
}
```

Here is the overview:

- One needs to bring in the CUDA defines:

  ```
  #include <cuda.h>
  ```

- The **main()** function runs on the CPU, as usual.

- The kernel function, **find1elt()** in this case, runs on the GPU, and is so denoted by the prefix **__global__**.

- The host code sets up space in the device memory via calls to **cudaMalloc()**, and transfers data from host to device or vice versa by calling **cudaMemcpy()**. The data on the device side is global to all threads.

- The host code launches the kernel via the lines

  ```
  dim3 dimGrid(n,1);   // n blocks in the grid
  dim3 dimBlock(1,1,1);   // 1 thread per block
  find1elt<<<dimGrid,dimBlock>>>(dm,drs,n);
  ```

- Each thread executes the kernel, working on a different row of the shared input matrix, and writing its result to a different element of the shared output array.

Other than the host/device distinction, the above description sounds very much like ordinary threaded programming. There is a major departure from the ordinary, though, in the structure of the threads.

Threads on the GPU are broken down into *blocks*, with the totality of all blocks being called the *grid*. In a kernel launch, we must tell the hardware how many blocks our grid is to have, and how many threads each block will have. Our code

```
dim3 dimGrid(n,1);   // n blocks in the grid
dim3 dimBlock(1,1,1);   // 1 thread per block
```

specifies n blocks per grid and one thread per block. One can also impose imaginary two- and three-dimensional structures on the grid and blocks, to be explained below; in the above code, the 1 in **dimGrid(n,1)** and the latter two 1s in **dimBlock(1,1,1)** here basically decline to use this feature. The advantage of having more than one thread per block will be discussed below. In this simple code, we have a separate thread for each row of the matrix.

In the kernel code itself, the line

```
int rownum = blockIdx.x;
```

determines which matrix row this particular thread will handle, as follows. Each block and thread has an ID, stored in programmer-accessible structs **blockIdx** and **threadIdx**, consisting of the block ID within the grid, and the thread ID within the block. Since in our case we've set up only one thread per block, the block ID is effectively the thread ID.

The **.x** field refers to first coordinate in the block ID. The "coordinates" of a block within the grid, and of a thread within a block, are merely abstractions. If for instance one is programming computation of heat flow across a two-dimensional slab, the programmer may find it clearer to use two-dimensional IDs for the threads. *But this does not correspond to any physical arrangement in the hardware.*

Some other points to mention:

- One compiles such code using **nvcc** in the CUDA toolkit, e.g.

  ```
  % nvcc -o rowsums RowSums.cu
  ```

  would produce an executable file **rowsums**. Note that **nvcc** is a wrapper for an underlying compiler such as **gcc**, so all the command-line options of the latter are retained. Adding the **-g** option will enable debugging, for instance (though only on the host).[1]

  The standard source file name suffix for CUDA is **.cu**. This can be overridden via the **-x** flag, e.g.

  ```
  % nvcc -x cu -o rowsums RowSums.c
  ```

  A very important option is to specify the target GPU architecture. Since the NVIDIA family has evolved over time, one needs to tell the compiler to prepare code for a certain family member or higher. This is specified with the **-arch** command line flag, e.g.

  ```
  nvcc x.cu -arch=sm_11
  ```

  tells the compiler that we need code for *compute capability* (an NVIDIA term) of at least 1.1, say because we call the CUDA function **atomicAdd()**.

- Another issue regarding GPU architecture is that earlier models did not support the **double** type. At least on my systems, though, **nvcc** automatically downgrades your **double** variables to **float**, with a warning message. Note that even if your GPU supports **double**, use of that type will slow down computation.

- Kernels can only have return type **void**. Thus a kernel must return its results through its arguments.

[1] Installing CUDA is beyond the scope of this book.

- Functions that will run on the device other than the kernel are denoted by the prefix **__device__**. These functions can have return values. They are called only by kernels or by other device functions.

By the way, a note on the innocuous-looking code

```
int *hm, // host matrix
    *dm, // device matrix
```

Here obviously **hm** is a memory address on the CPU. But in spite of the comment, **dm** is not a memory address on the GPU. Instead, **dm** will point to a C **struct** on the CPU, one of whose fields is the address of our matrix on the GPU. And of course, the CPU and GPU address spaces are not related to each other.[2]

6.4.2 NVIDIA GPU Hardware Structure

Scorecards, get your scorecards here! You can't tell the players without a scorecard—classic cry of vendors at baseball games

Know thy enemy—Sun Tzu, *The Art of War*

The enormous computational potential of GPUs cannot be truly unlocked without an intimate understanding of the hardware. This of course is a fundamental truism in the parallel processing world, but it is acutely important for GPU programming. This section presents an overview of the hardware.[3]

6.4.2.1 Cores

A GPU consists of a large set of *streaming multiprocessors* (SMs). Since each SM is essentially a multicore machine in its own right, you might say the GPU is a multi-multicore machine. Though the SMs run independently, they share the same GPU global memory. On the other hand, unlike ordinary multicore systems, there are only very limited (and slow) facilities for barrier synchronization and the like.

[2]Later GPU models do allow for a unified view of the two spaces.

[3]Some readers may benefit from some pictures, many good sets of which are available on the Web, such as `http://cs.nyu.edu/courses/spring12/CSCI-GA.3033-012/lecture5.pdf`.

Each SM in turn consists of a number of *streaming processors* (SPs)—the individual cores. The cores run threads, as with ordinary cores, but threads in an SM run in lockstep, to be explained below.

It is important to understand the motivation for this SM/SP hierarchy: Two threads located in different SMs cannot synchronize with each other in the barrier sense. Though this sounds like a negative at first, it is actually a great advantage, as the independence of threads in separate SMs means that the hardware can run faster. So, if the CUDA application programmer can write his/her algorithm so as to have certain independent chunks, and those chunks can be assigned to different SMs (we'll see how, shortly), then that's a "win."

6.4.2.2 Threads

As we have seen, when you write a CUDA application program, you partition the threads into groups called blocks. The salient points are:

- The hardware will assign an entire block to a single SM, though several blocks can run in the same SM.

- Barrier synchronization *is* possible for threads within the same block.

- Threads in the same block can access a programmer-managed cache called, confusingly, *shared memory.*

- The hardware divides each block into *warps*, currently 32 threads to a warp.

- Thread scheduling is handled on a warp basis. When some cores become free, this will occur with a set of 32 of them. The hardware then finds a new warp of threads to run on these 32 cores.

- *All the threads in a warp run the code in lockstep.* During the machine instruction fetch cycle, the same instruction will be fetched for all of the threads in the warp. Then in the execution cycle, each thread will either execute that particular instruction or execute nothing. The execute-nothing case occurs in the case of branches; see below.

  This is the classical *single instruction, multiple data* (SIMD) pattern used in some early special-purpose computers such as the ILLIAC; here it is called *single instruction, multiple thread* (SIMT).

Knowing that the hardware works this way, the programmer controls the block size and the number of blocks, and in general writes the code to take advantage of how the hardware works.

6.4.2.3 The Problem of Thread Divergence

The SIMT nature of thread execution has major implications for performance. Consider what happens with if/then/else code. If some threads in a warp take the "then" branch and others go in the "else" direction, they cannot operate in lockstep. That means that some threads must wait while others execute. This renders the code at that point serial rather than parallel, a situation called *thread divergence.* As one CUDA Web tutorial points out, this can be a "performance killer." Threads in the same block but in different warps can diverge with no problem.

6.4.2.4 "OS in Hardware"

Each SM runs the threads on a timesharing basis, just like an operating system (OS). This timesharing is implemented in the hardware, though, not in software as in the OS case. (The reader may wish to review Section 2.6 before continuing.)

The "hardware OS" runs largely in analogy with an ordinary OS:

- A thread in an ordinary OS is given a fixed-length timeslice, so that threads take turns running. In a GPU's hardware OS, warps take turns running, with fixed-length timeslices.

- With an ordinary OS, if a thread reaches an input/output operation, the OS suspends the thread while I/O is pending, even if its turn is not up. The OS then runs some other thread instead, so as to avoid wasting CPU cycles during the long period of time needed for the I/O.

  With an SM, the analogous situation occurs when there is a long memory operation, to global memory. If a warp of threads needs to access global memory, the SM will schedule some other warp while the memory access is pending. (Even for a currently-executing warp, the hardware accesses memory only one half-warp at a time.)

The hardware support for threads is extremely good; a context switch from one warp to another takes very little time, quite a contrast to the OS case. Moreover, as noted above, the long latency of global memory may be solvable by having a lot of threads that the hardware can timeshare to hide that latency; while one warp is fetching data from memory, another warp can be executing, thus not losing time due to the long fetch delay. For these reasons, CUDA programmers typically employ a large number of threads, each of which does only a small amount of work—again, quite a contrast to something like OpenMP.

6.4.2.5 Grid Configuration Choices

In choosing the number of blocks and the number of threads per block, one typically knows the number of threads one wants (recall, this may be far more than the device can physically run at one time, due to the desire to ameliorate memory latency problems), so configuration mainly boils down to choosing the block size. This is a delicate art, again beyond the scope of this book, but here is an overview of the considerations:

- The device will have limits on the block size, number of threads on an SM, and so on. (See Section 6.4.2.9 below.)

- Given that scheduling is done on a warp basis, block size should be a multiple of the warp size, currently 32.

- One wants to utilize all the SMs. If one sets the block size too large, not all will be used, as a block cannot be split across SMs.

- As noted, barrier synchronization can be done effectively only at the block level. The larger the block, the more the barrier delay, so one might want smaller blocks.

- On the other hand, if one is using shared memory, this can only be done at the block level, and efficient use may indicate using a larger block.

- Two threads doing unrelated work, or the same work but with many if/elses, would cause a lot of thread divergence if they were in the same block. In some cases, it may be known in advance which threads will do the "ifs" and which will do the "elses," in which case they should be placed in different blocks if possible.

- A commonly-cited rule of thumb is to have between 128 and 256 threads per block.

6.4.2.6 Latency Hiding in GPUs

In our code example in Section 6.4.1, we had one thread per row of the matrix. That degree of fine-grained parallelism may surprise those who are used to classical shared-memory programming. Section 2.7 did note that it may be beneficial to have more threads than cores, due to cache effects and so on, but in the multicore setting this would mean just a *few* more. By contrast, in the GPU world, it's encouraged to have lots of threads,[4] in order to circumvent the memory latency problems in GPUs: If the GPU operating system senses that there may be quite a delay in the memory access needed by a given thread, that thread is suspended and another is found to run; by having a large number of threads, we ensure that the OS will succeed in finding a new thread for this. Here we are using *latency hiding* (Section 2.5).

Note too that while the previous paragraph spoke of the OS sensing that *a* thread faces a memory delay, that was an oversimplification. Since threads are scheduled in warps, if just one thread faces a memory delay, then the entire warp must wait.

On the other hand, this is actually a plus, as follows. The global memory in GPUs uses *low-order interleaving*, which means that consecutive memory addresses are physically stored in simultaneously accessible places. And furthermore, the memory is capable of *burst mode*, meaning that one can request accesses to several consecutive locations at once.

This means we can reap great benefits if we can design our code so that consecutive threads access consecutive locations in memory. In that case, you can see why the NVIDIA designers were wise to schedule thread execution in warp groups, giving us excellent latency hiding.

6.4.2.7 Shared Memory

As noted earlier, the GPU has a small amount of *shared memory*, the term *shared* meaning that threads in a block share that storage. Access to shared memory is both low-latency and high-bandwidth, compared to access of global memory, which is off-chip. As noted, the size is small, so the programmer must anticipate what data, if any, is likely at any given time to be accessed repeatedly by the code. If data exists, the code can copy it from global memory to shared memory, and access the latter, for a performance win. In essence, shared memory is a programmer-managed cache.

[4] There are limits, however, and a request to set up too many threads may fail.

The shared memory is allocated in the kernel launch, as a third configuration parameter, e.g.

```
sieve<<<dimGrid, dimBlock, psize>>>(dprimes, n, nth);
```

It needs declaration within a kernel, e.g.

```
extern __shared__ int sprimes[];
```

6.4.2.8 More Hardware Details

Details of the use of shared memory are beyond the scope of this book. Indeed, there is so much more than what could be comfortably included here.

All this illustrates why our discussion in Section 6.2 recommended that most users either (a) settle for writing "pretty fast but simple" CUDA code, and/or (b) rely mainly on libraries of either preoptimized CUDA code or R code that interfaces to high-quality CUDA code.

6.4.2.9 Resource Limitations

Any CUDA device has limits on the number of blocks, threads per block and so on. For safety, calls to **cudaMalloc()** should be accompanied by error checking, something like

```
if (cudaSuccess != cudaMalloc(...)) {...}
```

Here is code that prints out some selected resource limits:

```
#include <cuda.h>
#include <stdio.h>

int main()
{
   cudaDeviceProp Props;
   cudaGetDeviceProperties(&Props,0);

   printf("shared mem: %d)\n",
      Props.sharedMemPerBlock);
   printf("max threads/block: %d\n",
      Props.maxThreadsPerBlock);
   printf("max blocks: %d\n",Props.maxGridSize[0]);
```

```
   printf("total Const mem: %d\n",Props.totalConstMem);
}
```

I compiled and ran the program on my machine:

```
% nvcc -o cudaprops CUDAProperties.cu
% cudaprops
shared mem: 49152)
max threads/block: 1024
max blocks: 65535
total Const mem: 65536
```

Consult the documentation for **cudaGetDeviceProperties()** to determine other resource limits.

6.5 Example: Mutual Inlinks Problem

Once again, our example involves our mutual Web outlinks problem, first introduced in Section 1.4. Here is one approach to the calculation on a GPU, but with the same change made in Section 5.9.1: We now look at *inbound* links. Thus we are looking for matches of 1s between pairs of columns of the adjacency matrix, as opposed to rows in the outlinks examples. (Or equivalently, suppose we are interested in outlinks but are storing the matrix in transpose form.)

The reason for this change is that the example will illustrate the principle of GPU latency hiding discussed in Section 6.4.2.6. Consider threads 3 and 4, say. In the "i" (i.e., outermost) loop in the code below. these two threads will be processing consecutive columns of the matrix. Since the GPU is running the two threads in lockstep, and because storage is in row-major order, thread 4 will always be accessing an element of the matrix in the same row as thread 3, but with the two accesses being to adjacent elements. Thus we can really take advantage of burst mode in the memory chips.

6.5.1 The Code

```
// MutIn.cu:  finds mean number of mutual inlinks,
// among all pairs of Web sites in our set; in
// checking (i,j) pairs, thread k will handle all i
```

```
// such that i mod totth = k, where totth is
// the number of threads

// usage:
//
//    mutin numvertices numblocks

#include <cuda.h>
#include <stdio.h>
#include <stdlib.h>

// block size is hard coded as 192
#define BLOCKSIZE 192

// kernel:  processes all pairs assigned to
// a given thread
__global__ void procpairs(int *m, int *tot, int n)
{
   // total number of threads =
   // number of blocks * block size
   int totth = gridDim.x * BLOCKSIZE,
       // my thread number
       me = blockIdx.x * blockDim.x + threadIdx.x;
   int i,j,k,sum = 0;
   // various columns i
   for (i = me; i < n; i += totth) {
      for (j = i+1; j < n; j++) {  // all columns j > i
         for (k = 0; k < n; k++)
            sum += m[n*k+i] * m[n*k+j];
      }
   }
   atomicAdd(tot,sum);
}

int main(int argc, char **argv)
{  int n = atoi(argv[1]),  // number of vertices
       nblk = atoi(argv[2]);  // number of blocks
    // the usual initializations
    int *hm, // host matrix
        *dm, // device matrix
        htot, // host grand total
        *dtot; // device grand total
    int msize = n * n * sizeof(int);
```

```
   hm = (int *) malloc(msize);
   // as a test, fill matrix with random 1s and 0s
   int i,j;
   for (i = 0; i < n; i++) {
      hm[n*i+i] = 0;
      for (j = 0; j < n; j++) {
         if (j != i) hm[i*n+j] = rand() % 2;
      }
   }
   // more of the usual initializations
   cudaMalloc((void **)&dm, msize);
   // copy host matrix to device matrix
   cudaMemcpy(dm,hm, msize, cudaMemcpyHostToDevice);
   htot = 0;
   // set up device total and initialize it
   cudaMalloc((void **)&dtot, sizeof(int));
   cudaMemcpy(dtot,&htot, sizeof(int),
      cudaMemcpyHostToDevice);
   // OK, ready to launch kernel, so configure grid
   dim3 dimGrid(nblk,1);
   dim3 dimBlock(BLOCKSIZE,1,1);
   // launch the kernel
   procpairs<<<dimGrid, dimBlock>>>(dm, dtot,n);
   // wait for kernel to finish
   cudaThreadSynchronize();
   // copy total from device to host
   cudaMemcpy(&htot, dtot, sizeof(int),
      cudaMemcpyDeviceToHost);
   // check results
   if (n <= 15) {
      for (i = 0; i < n; i++) {
         for (j = 0; j < n; j++)
            printf("%d ",hm[n*i+j]);
         printf("\n");
      }
   }
   printf("mean = %f\n",htot/(float)((n*(n-1))/2));
   // clean up
   free(hm);
   cudaFree(dm);
   cudaFree(dtot);
}
```

By now the reader will immediately recognize that the reason for staggering the columns is that a given thread handles is to achieve load balance. Also, we've continued to calculate the "dot product" between each pair of columns, to avoid thread divergence. The only new material, a call to **atomicAdd()** is discussed in Section 6.6.

6.5.2 Timing Experiments

The code was run on a machine with a Geforce 285 GPU, for various numbers of blocks, with a comparison to ordinary CPU code. Here are the results, with times in seconds:

blocks	time
CPU	97.26
4	4.88
8	3.17
16	2.48
32	2.36

So, first of all, we see that using the GPU brought us a dramatic gain! Note, though, that while GPUs work quite well for certain applications, they do poorly on others. Any algorithm that necessarily requires a lot of if-then-else operations, for instance, is a poor candidate for GPUs, and the same holds if the algorithm needs a considerable number of synchronization operations. Even in our Web link example here, the corresponding speedup was far more modest for the outlinks version (not shown here), due to the memory latency issues we've discussed.

Second, the number of blocks does matter. But since our block size was 192, we can only use about 1750/192 blocks, so there was no point in going beyond 32 blocks.

6.6 Synchronization on GPUs

The call

```
atomicAdd(tot,sum);
```

is similar to the material in Section 5.12. There it was mentioned that in some CPU architectures it is possible to do operations like *atomic add*, in which a single machine instruction increments a shared sum, with the operation being atomic but not requiring lock variables. The latter aspect can substantially enhance performance.

The NVIDIA GPUs (except for the earliest models) do offer several atomic operations like this, such as the one we've used above.[5] Others include **atomicExch()** (exchange the two operands), **atomicCAS()** (if the first operand equals the second, replace the first by the third), **atomicMin()**, **atomicMax()**, **atomicAnd()**, **atomicOr()**, and so on.

In compiling code with these operations, we must warn the compiler that it needs to produce an executable that runs on Model 1.1 and above:

```
% nvcc -o mutin MutIn.cu -arch=sm_11
```

These operations look good, but appearances can be deceiving, in this case masking the fact that these operations are extremely slow. For example, though a barrier could in principle be constructed from the atomic operations, its overhead would be quite high. In earlier models that delay was near a microsecond, and though that problem has been ameliorated in more recent models, implementing a barrier in this manner would not be not much faster than attaining interblock synchronization by returning to the host and calling **cudaThreadSynchronize()** there. The latter *is* a possible way to implement a barrier, since global memory stays intact in between kernel calls, but again, it would be slow.

NVIDIA does offer barrier synchronization at the block level, via a call to **__syncthreads()**. This is reasonably efficient, but it still leaves us short of an efficient way to do a global barrier operation, across the entire GPU.

The overall outcome from all of this is that algorithms that rely heavily on barriers may not run well on GPUs.

6.6.1 Data in Global Memory Is Persistent

Data in global memory persists through the life of the program. In other words, if our program does

```
allocate space for an array dx in GPU global memory
copy a host array hx to dx
call some kernel that uses (and possibly modifies) dx
do some more computation
call some kernel that uses (and possibly modifies) dx
```

[5]Note that the first argument must be an address on the GPU, which it is here.

then in that second kernel execution, **dx** will still contain whatever data it had at the end of the first kernel call.

The above scenario occurs in many GPU-based applications. For example, consider iterative algorithms. As noted in Section 6.6 it is difficult to have a barrier operation across blocks, and it may be easier just to return to the CPU after each iteration, in essence using **cudaThreadSynchronize()** to implement a barrier. Or, the algorithm itself may not be iterative, but our data may be too large to fit in GPU memory, necessitating doing the computation one chunk at a time, with a separate kernel call for each chunk.

In such scenarios, if we had to keep copying back and forth between **hx** and **dx**, we may incur very significant delays. Thus we should exploit the fact that data in global memory does persist across kernel calls.[6]

Some GPU libraries for R, such as **gputools**, do not exploit this persistence of data in global memory. However, the **gmatrix** package on CRAN does do this, as does **RCUDA**.

6.7 R and GPUs

Given the numerical nature of most R applications, it is natural that packages have been developed for interface between R and GPUs, yet another example of the "R+X" notion introduced in Section 1.1. A good place to check what is currently available in this regard is the CRAN Task View, "High-Performance and Parallel Computing with R" (`http://cran.r-project.org/web/views/HighPerformanceComputing.html`). This is especially true for matrix operations, which by virtue of their regular pattern makes them generally well-suited to GPU computation.

Examples using R packages for GPUs will often appear in the succeeding chapters, but let's do one here first:

6.7.1 Example: Parallel Distance Computation

The **gputools** package is probably the most commonly-used GPU package for R. It consists mainly of linear algebra operations, but let's take a look at what it does for distance computation.

[6]By contrast, for instance, data in shared memory lives only during the given kernel call.

In Section 3.9, we used **snow** to parallelize computation of distances of rows in one matrix to rows in another. The base R function **dist()** calculates distances of rows in a single matrix to those in the *same* matrix. The **gputools** function **gpuDist()** does this intramatrix computation too, but on a GPU. In their simplest call forms, both functions have a single argument, the matrix.

Again, the very regular nature of the computation here should allow the GPU to bring us major speedups. This is indeed the case.

I installed the package as usual:[7]

```
> install.packages("gputools","~/Pub/Rlib")
```

I filled $n \times n$ matrices with U(0,1) data, for various n, comparing run times. Here are the results, in seconds:

n	dist()	gpuDist(
1000	3.671	0.258
2500	103.219	3.220
5000	609.271	error

The speedups in the first two cases are quite impressive, but an execution error occurred in the last case, with a message, "the launch timed out and was terminated." The problem here was that the GPU was being used both for computation and for the ordinary graphics screen of the computer housing the GPU. One can disable the timeout if one owns the machine, or better, purchase a second GPU for computation only. GPU work is not easy...

6.8 The Intel Xeon Phi Chip

GPUs are just one kind of *accelerator chip*. Others exist, and in fact go back to the beginning of the PC era in the 1980s, when one could purchase a floating-point hardware coprocessor.

The huge success of the NVIDIA GPU family did not go unnoticed by Intel. In 2013 Intel released the Xeon Phi chip, which had been under development for several years. At this writing (Spring 2014), NVIDIA has

[7]If you have CUDA installed in a nonstandard location, you'll need to download the **gputools** source and then build using **R CMD INSTALL**, specifiying the locations of the library and include files. See the **INSTALL** file that is included in the package. During execution and use, be sure your environment variable **LD_LIBRARY_PATH** includes the CUDA library.

a large head start in this market, so it is unclear how well the Intel chip will do in the coming years.

The Xeon Phi features 60 cores, each 4-way hyperthreaded, thus with a theoretical level parallelism level of 240. This number is on the low end of NVIDIA chips. But on the other hand, the Intel chip is much easier to program, as it has much more of a classic multicore design. One can run OpenMP, MPI and so on. As with GPUs, though, the bandwidth and latency for data transfers between the CPU and the accelerator chip can be a major issue.

It must again be kept in mind that *while the NVIDIA chips can attain exceptionally good performance on certain applications, they perform poorly on others.* A number of analysts have done timing tests, and there are indeed applications for which the Intel chip seems to do better than, for instance, the NVIDIA Tesla series, in spite of having fewer cores.

6.9 Further Reading

CUDA has many more features than those discussed in the brief treatment here. A number of technical books present much more detail, such as *Professional CUDA C Programming*, by John Cheng *et al*, Wiley, 2014.

Effective CUDA programming requires inclusion of many "tweaks" to the code, motivated by very delicate considerations of memory access times and so on. The author highly recommends the excellent case study, *Histogram Calculation in CUDA*, by Victor Podlozhnyuk of NVIDIA, 2007, at `http://developer.download.nvidia.com/compute/cuda/1.1-Beta/x86_website/projects/histogram64/doc/histogram.pdf`.

Chapter 7

Thrust and Rth

As discussed in Section 6.2, GPU programming is difficult to do at true efficiency, so it was recommended to make use of GPU code libraries whenever possible. One such library is Thrust, to be presented in this chapter.

In turn, Thrust programming itself requires good facility with C++, especially some of the more esoteric features. An alternative is the **Rth** package, which gives R programmers access to certain algorithms written in Thrust, *without those programmers needing to know Thrust/C/C++/CUDA*; this package too will be presented here.

Even more valuable is the fact that Thrust, and thus **Rth**, can be used not only on GPU-equipped machines, but also on multicore systems! In fact, Thrust can be directed to produce OpenMP or Thread Building Blocks (TBB code, Section 5.11) code.

7.1 Hedging One's Bets

It is very unclear how the price-performance tradeoff for shared-memory systems will evolve in the coming years—multicore, GPU, accelerators like the Intel Xeon Phi, or whatever else may be in store for us. On the other hand, as discussed in Section 4.2, the shared memory paradigm is attractive, especially for certain types of applications, and it would be desirable to write code that works on various types of shared memory hardware.

Several software systems have been developed to "hedge one's bets" in this sense. OpenCL, for instance, is an extension of C/C++ that is designed for such *heterogeneous computing.* Two such "bet hedging" systems, Thrust and **Rth**, will be presented here.

7.2 Thrust Overview

Thrust was developed by NVIDIA, maker of the graphics cards that run CUDA. It consists of a collection of C++ templates, modeled after the C++ Standard Template Library (STL). One big advantage of the template approach is that no special compiler is needed; Thrust is simply a set of "#include" files.

In one sense, Thrust may be regarded as a higher-level way to develop GPU code, avoiding the tedium of details that arise when programming GPUs. But more important, as noted above, Thrust does enable heterogeneous computing, in the sense that it can produce different versions of machine code to run on different platforms.

When one compiles code using Thrust, one can choose the *backend*, be it GPU code or multicore. In the latter case one can currently choose OpenMP or TBB for multicore machines.

In other words, Thrust allows when to write a single piece of code that can run either on a GPU or on a classic multicore machine. The code of course will not be optimized, but in many cases one will attain reasonable speed for this diversity of hardware types.

7.3 Rth

As will be seen below, even though Thrust is designed to ease the task of GPU programming, coding in Thrust is somewhat difficult. C++ template code can become quite intricate and abstract if you are not used to something like the STL (or even if you do have STL experience).

Thus once again it is desirable to have R libraries available that interface to a lower-level language, the "R+X" concept mentioned in Section 1.1.2. Drew Schmidt and I have developed **Rth** (`https://github.com/Rth-org/Rth`), an R package that builds on Thrust. **Rth** implements a number of R-callable basic operations in Thrust.

The goal is to give the R programmer the advantages of Thrust, without having to program in Thrust (let alone C++, CUDA or OpenMP) him or herself.

For instance, **Rth** provides a parallel sort, which is really just an R wrapper for the Thrust sort. (See Section 10.5.) Again, this is a versatile sort, in that it can take advantage of both GPUs and multicore machines.

7.4 Skipping the C++

C++ is increasingly becoming an important element of the R programmer's toolkit. However, readers who wish to put aside C++ for now can skip the Thrust programming material, and go directly to **Rth**, Section 7.6.

7.5 Example: Finding Quantiles

Although the initial learning curve is a bit steep, Thrust programming is straightforward. Below is Thrust code to extract every k^{th} element of an array. It thus finds the $i \cdot k/n$ quantiles of the data, $i = 1, 2, ...$ Note that though the "device" is the GPU if that is the backend, in the multicore case it is simply shared memory.[1]

7.5.1 The Code

There are of course many ways to write code for a given application, and that certainly is the case here. The approach shown below is here mainly to illustrate how Thrust works, rather than being claimed an optimal implementation.

```
// Quantiles.cpp, Thrust example

// calculate every k-th element in given numbers,
// going from smallest to largest; k obtained from
// command line and fed into the ismultk() functor

// these are the ik/n * 100 percentiles, i = 1, 2, ...

#include <stdio.h>
```

[1] If one is only going to use multicore backends, it's better not to copy to a device, to avoid the overhead in copying potentially very large objects.

```
#include <thrust/device_vector.h>
#include <thrust/sort.h>
#include <thrust/sequence.h>
#include <thrust/copy.h>

// functor
struct ismultk {
   const int increm;  // k in above comments
   // get k from call
   ismultk(int _increm): increm(_increm) {}
   __device__ bool operator()(const int i)
   {  return i != 0 && (i % increm) == 0;
   }
};

int main(int argc, char **argv)
{  int x[15] =
      {6,12,5,13,3,5,4,5,8,88,1,11,9,22,168};
   int n=15;
   // create int vector dx on the device, init. to
   // x[0], x[1], ..., x[n-1]
   thrust::device_vector<int> dx(x,x+n);
   // sort dx in-place
   thrust::sort(dx.begin(),dx.end());
   // create a vector seq of length n
   thrust::device_vector<int> seq(n);
   // fill seq with 0,1,2,...n-1
   thrust::sequence(seq.begin(),seq.end(),0);
   // set up space to store our quantiles
   thrust::device_vector<int> out(n);
   // obtain k from command line
   int incr = atoi(argv[1]);
   // for each i in seq, call ismultk() on this i,
   // and if get a true result, put dx[i] into out;
   // return pointer to (one element past) the end
   // of this output array
   thrust::device_vector<int>::iterator newend =
      thrust::copy_if(dx.begin(),dx.end(),seq.begin(),
         out.begin(), ismultk(incr));
   // print results
   thrust::copy(out.begin(), newend,
      std::ostream_iterator<int>(std::cout, " "));
   std::cout << "\n"; }
```

7.5.2 Compilation and Timings

As mentioned, a big advantage of Thrust is that one doesn't need a special compiler, as it consists only of "include" files. So, to compile a Thrust application for a multicore backend, say for OpenMP, one can use **gcc** or any other OpenMP-enabled compiler. Of course, for a GPU backend, one uses **nvcc**.

The Thrust library is included with CUDA. But the multicore machine used in many of the examples in this book does not have an NVIDIA GPU, so I downloaded Thrust from `http://thrust.github.com/`, and unzipped the package in a subdirectory I chose to name **Thrust** in my home directory. The unzip operation produced the further subdirectory **Thrust/thrust**, in which the **.h** files reside.

To compile the above source, **Quantiles.cpp**, for an OpenMP backend, I typed

```
g++ -o quants Quantiles.cpp -fopenmp \
   -DTHRUST_DEVICE_SYSTEM=THRUST_DEVICE_SYSTEM_OMP \
   -lgomp -I/home/matloff/Thrust -g
```

For TBB, I used

```
g++ -o quants Quantiles.cpp -g \
   -DTHRUST_DEVICE_SYSTEM=THRUST_DEVICE_SYSTEM_TBB \
   -ltbb -I/home/matloff/Thrust \
   -I/home/matloff/Pub/TBB/include \
   -L/home/matloff/Pub/TBB/lib/tbb
```

I later tried it on a GPU on a different machine, typing

```
nvcc -x cu -o quants Quantiles.cpp -g
```

(No need to state where Thrust is, as it is built in to CUDA.)

7.5.3 Code Analysis

In Thrust, one works with vectors rather than arrays, in the sense that vectors are objects, very much like in R. They thus have built-in methods, leading to expressions such as **dx.begin()**, a function call that returns the location of the start of **dx**. Similarly, **dx.size()** tells us the length of **dx**, **dx.end()** points to the location one element past the end of **dx** and so on.

Note that though **dx.begin()** acts like a pointer, it is called an *iterator*, with broader powers than those of a pointer.

Most of the vector operations in the code are described in the comments. But where things get interesting is the struct **ismultk**, which though an ordinary C struct, is known as a *functor* in C++ terminology.

A functor is a C++ mechanism to produce a callable function, largely similar in goal to using a pointer to a function but with the added notion of saved state. This is done by turning a C++ struct or class object into a callable function. Since structs and classes can have member variables, we can store needed data in them, and that is what distinguishes functors from function pointers. Consider the code:

```
thrust::device_vector<int>::iterator newend =
   thrust::copy_if(dx.begin(),dx.end(),seq.begin(),
      out.begin(),ismultk(incr));
```

The key here is the function **copy_if()**, which as the name implies copies all elements of an input vector (**dx** here) that satisfy a certain predicate. The latter role is played by **ismultk**, to be explained shortly, with the help of a *stencil*, in this case the vector **seq**.

The output is stored here in **out**, but not all of that vector will be filled. Thus the return value of **copy_if()**, assigned here to **newend**, is used to inform us as to where the output actually ends.

Now let's look at **ismultk** ("is a multiple of k"):

```
struct ismultk {
   const int increm;  // k in above comments
   // get k from call
   ismultk(int _increm): increm(_increm) {}
   __device__ bool operator()(const int i)
   {  return i != 0 && (i % increm) == 0;
   }
};
```

The keyword **operator** here tells the compiler that **ismultk** will serve as a function. At the same time, it is also a struct, containing the data **increm**, which as the comments note, is "k" in our description "take every k^{th} element of the array."

Now, how is that function called? Let's look at the **copy_if()** call examined above:

```
thrust::copy_if(dx.begin(),dx.end(),seq.begin(),
   out.begin(), ismultk(incr));
```

Inside that call is another call! It is in the code

```
ismultk(incr)
```

and what that call does is instantiate the struct **ismultk**. In other words, that call returns a struct of type **ismultk**, with the member variable **increm** in that struct being set to **incr**. That assignment is done in the line,

```
ismultk(int _increm): increm(_increm) {}
```

In other words: The inner call to **ismultk()**

```
thrust::copy_if(dx.begin(),dx.end(),seq.begin(),
   out.begin(), ismultk(incr));
```

returns a function, whose body is

```
__device__ bool operator()(const int i)
{  return i != 0 && (i % increm) == 0;
}
```

What **copy_if()** does is apply that function to all values **i** in the stencil, where the latter consists of 0,1,2,...,n-1.

So all of this is a roundabout way of copying **dx[i]** to **out** for i = k, 2k, 3k, ..., with ik not to exceed n-1. This is exactly what we want.

The word *roundabout* above is apt, and arguably is typical of Thrust (and, for that matter, of the C++ STL). But we do get hardware generality from that effect, with our code being applicable both to GPUs and multicore platforms.

Important note: If you plan to write Thrust code on your own, or any other code that uses functors, C++11 *lambda functions* can simplify things a lot. See Section 11.6.4 for an example.

7.6 Introduction to Rth

As noted, download the package from `https://github.com/Rth-org/Rth`. For installation instructions, see the **INSTALL** file included in the package. Further information is available at the project home page, `http://heather.cs.ucdavis.edu/~matloff/rth.html`.

As an example, let's compute standard (Pearson product-moment) correlation, comparing the run time to R's built-in **cor.test()** function, with two and then eight threads on a multicore backend:

```
> n <- 100000000
> tmp <- matrix(runif(2*n),ncol=2)
> x <- tmp[,1]
> y <- x + tmp[,2]
> system.time(c1 <- cor.test(x,y))
   user  system elapsed
 13.150   1.166  14.333
> c1$estimate
      cor
0.7071664
> system.time(c2 <- rthpearson(x,y,2))
   user  system elapsed
  2.843   0.514   2.471
> c2
[1] 0.7071664
> system.time(c2 <- rthpearson(x,y,8))
   user  system elapsed
  3.540   0.529   1.792
```

Similar times emerged in repeat runs. This is a better-than-linear speedup, probably due to a difference in algorithms.

As mentioned earlier, R programmers can use **Rth** without knowledge of C++/CUDA/OpenMP/Thrust. But for those who wish to develop their own **Rth** functions, here is how **rthpearson()** looks inside:

```
// Rth implementation of Pearson product-moment
// correlation

// single-pass, subject to increased roundoff error

#include <thrust/device_vector.h>
#include <thrust/inner_product.h>
```

```
#include <math.h>

#include <Rcpp.h>
#include "backend.h"

typedef thrust::device_vector<int> intvec;
typedef thrust::device_vector<double> doublevec;

RcppExport SEXP rthpearson(SEXP x, SEXP y,
   SEXP nthreads)
{
   Rcpp::NumericVector xa(x);
   Rcpp::NumericVector ya(y);
   int n = xa.size();
   doublevec dx(xa.begin(),xa.end());
   doublevec dy(ya.begin(),ya.end());
   double zero = (double) 0.0;

   #if RTH_OMP
   omp_set_num_threads(INT(nthreads));
   #elif RTH_TBB
   tbb::task_scheduler_init init(INT(nthreads));
   #endif

   double xy =
      thrust::inner_product(dx.begin(),dx.end(),
         dy.begin(),zero);
   double x2 =
      thrust::inner_product(dx.begin(),dx.end(),
         dx.begin(),zero);
   double y2 =
      thrust::inner_product(dy.begin(),dy.end(),
         dy.begin(),zero);
   double xt =
      thrust::reduce(dx.begin(),dx.end());
   double yt =
      thrust::reduce(dy.begin(),dy.end());
   double xm = xt/n, ym = yt/n;
   double xsd = sqrt(x2/n - xm*xm);
   double ysd = sqrt(y2/n - ym*ym);
   double cor = (xy/n - xm*ym) / (xsd*ysd);
   return Rcpp::wrap(cor);
}
```

Here the Thrust function **inner_product()** performs the "dot product" $\sum_{i=1}^{n} X_i Y_i$, and Thrust's **reduce()** does a reduction as with constructs of a similar name in OpenMP and **Rmpi**; the default operation is addition.

Chapter 8

The Message Passing Paradigm

The scatter-gather paradigm we've seen in earlier examples with **snow** works well for many problems, but it can be confining. This chapter will present a more general approach.

Instead of a situation in which the workers communicate only with the manager, as in for instance in **snow**, think now of allowing the workers to send messages to each other as well. This general case is known as the *message passing* paradigm, the subject of this chapter.

8.1 Message Passing Overview

A message-passing package will have some kind of **send()** and **receive()** functions for its basic operations, along with variants such as broadcasting messages to all processes. In addition, there may be functions for other operations, such as:

- *Scatter/gather* (Section 1.4.4).
- *Reduction*, similar to R's **Reduce()** function.
- *Remote procedure call*, in which one process triggers a function call at another process, similar to **clusterCall()** in **snow**.

The most popular C-level package for message passing is the Message Passing Interface (MPI), a collection of routines callable from C/C++.[1] Professor Hao Yu of the University of Western Ontario wrote an R package, **Rmpi**, that interfaces R to MPI, as well as adding a number of R-specific functions. **Rmpi** will be our focus in this chapter. (Two other popular message-passing packages, PVM and 0MQ, also have had R interfaces developed for them, **Rpvm** and **Rzmq**, as well as a very promising new R interface to MPI, **pdbR**.)

So with **Rmpi**, we might have, say, eight machines in our cluster. When we run **Rmpi** from one machine, that will then start up R processes on each of the other machines. This is the same as what happens when we use **snow** on a physical cluster, where for example the call **makeCluster(8)**, causes there to be 8 R processes created on the manager's machine. The various processes will occasionally exchange data, via calls to **Rmpi** functions, in order to run the given application in parallel. Again, this is the same as for **snow**, but here the workers can directly exchange data with each other.

We'll cover a specific example shortly. But first, let's follow up on the discussion of Section 2.5, and note the special issues that arise with message passing code. (The reader may wish to review the concepts of latency and bandwidth in that section before continuing.)

8.2 The Cluster Model

Message passing is a software/algorithmic notion, and thus does not imply any special structure of the underlying hardware platform. However, message passing is typically thought of as being run on a cluster, i.e., a network of independent standalone machines, each having its own processor and memory. So, although MPI and **Rmpi** can be run on a multicore machine, which is quite common, the mental model is still the cluster. We'll assume this situation throughout.

In a small business or university computing lab, for instance, one may have a number of PCs, connected by a network. Though each PC runs independently of the others, one can use the network to pass messages among the PCs, thus forming a parallel processing system. In a dedicated cluster, the nodes are typically not even full PCs; since the nodes are not used as independent, general-purpose computers, one dispenses with the keyboards and monitors, and places multiple PCs on the same rack.

[1] For an introduction to MPI, see my online book, *Programming on Parallel Machines*, http://heather.cs.ucdavis.edu/parprocbook.

8.3 Performance Issues

Recall the discussion of network infrastructures in Section 2.4. The network is, literally, the weakest link, meaning the major source of slowdown. In data science applications, this delay can be especially acute, as copying large amounts of data incurs a large time penalty.

A good cluster will typically have a fancier network than the standard Ethernet used in an office or lab. An example is InfiniBand. In this technology, the single communications channel is replaced by multiple point-to-point links, connected by switches.

The fact that there are multiple links means that potential bandwidth is greatly increased, and contention for a given link is reduced. InfiniBand also strives for low latency.

Note, though, that even with InfiniBand, latency is on the order of a microsecond, i.e., a millionth of a second. Since CPU clock speeds are typically more than a gigahertz, i.e., CPUs are capable of billions of operations per second, even InfiniBand network latency presents considerable overhead.

One way of reducing the overhead arising from the network system software is to use *remote direct memory access* (RDMA), which involves both nonstandard hardware and software. The name derives from the Direct Memory Access devices that are common in even personal computers today.

When reading from a fast disk, for instance, DMA can bypass the "middleman," the CPU, and write directly to memory, a significant speedup. (DMA devices in fact are special-purpose CPUs in their own right, designed to copy data directly between an input-output device and memory.) Disk writes can be done the same way.

With RDMA, we bypass a different kind of middleman, in this case the network protocol stack in the operating system. When reading a message arriving from the network, RDMA deposits the message directly into the memory used by our program, bypassing the layers in the OS that handle ordinary network traffic.

8.4 Rmpi

As noted, **Rmpi** is an R interface to the famous MPI protocol, the latter normally being accessed via C, C++ or FORTRAN. MPI consists of dozens of functions callable from user programs.

Note that MPI also provides network services beyond simply sending and receiving messages. An important point is that it enforces message order. If say, messages A and B are sent from process 8 to process 3 in that order, then the program at process 3 will receive them in that order. A call at process 3 to receive from process 8 will receive A first, with B not being processed until the second such call.[2]

This makes the logic in your application code much easier to write. Indeed, if you are a beginner in the parallel processing world, keep this foremost in mind. Code that makes things happen in the wrong order (among the various processes) is one of the most common causes of bugs in parallel programming.

In addition, MPI allows the programmer to define several different kinds of messages. One might make a call, for instance, that says in essence, "read the next message of type 2 from process 8," or even "read the next message of type 2 from any process."

Rmpi provides the R programmer with access to such operations, and also provides some new R-specific messaging operations. It is a very rich package, and we can only provide a small introduction here.

8.4.1 Installation and Execution

With all that power comes complexity. **Rmpi** can be tricky to install—and even to launch—with various platform dependencies to deal with, such as those related to different flavors of MPI. Since there are too many possible scenarios, I will simply discuss an example setup.

On a certain Linux machine, I had installed MPI in a directory **/home/-matloff/Pub/MPICH**, and then installed **Rmpi** as follows. After downloading the source package from CRAN, I ran

```
$ R CMD INSTALL -l ~/Pub/Rlib Rm*z \\
--configure-args="--with-mpi=/home/matloff/Pub/MPICH \\
--with-Rmpi-type=MPICH"
```

Here I stored **Rmpi** in the directory **/home/matloff/Pub/Rlib**.

The more difficult part was getting the package to actually run. With some versions of MPI, the function **mpi.spawn.Rslaves()**, intended as a standard way for the manager process to launch workers doesn't work.[3]

[2]This assumes that the calls do not specify message type, discussed below.

[3]**Rmpi** uses *master/slave* terminology instead of *manager/worker*, and has been the subject of some criticism on this point.

Instead, I did the following setup to enable running **Rmpi**, making use of a file **Rprofile** that comes with the **Rmpi** package for this purpose:

```
$ mkdir ~/MyRmpi
$ cd ~/MyRmpi
# make copy of R
$ cp /usr/bin/R Rmpi
# make it runnable
$ chmod u+x Rmpi
# edit my shell startup file (not shown):
#     place Rmpi file in my execution path
#     add /home/matloff/Pub/MPICH/lib
#     to LD_LIBRARY_PATH
# edit Rmpi file (not shown):
#     after "export R_HOME", insert
#     R_PROFILE=/home/matloff/MyRmpi/Rprofile;
#     export R_PROFILE
$ cp ~/Pub/Rlib/Rmpi/Rprofile .
# edit Rprofile (not shown):
#     insert at top
#     ".libPaths('/home/matloff/Pub/Rlib')"
# test:
# set up to run all processes on local machine
$ echo "localhost" > hosts
# run MPI on the Rmpi file, with 3 processes
$ mpiexec -f hosts -n 3 Rmpi --no-save -q
# R now running, with Rmpi loaded, mgr and 2 wrkrs
> mpi.comm.size()  # number of processes, should be 3
# have the 2 workers run sum(), one on 1:3 and
# the other on 4:5
> mpi.apply(list(1:3, 4:5),sum)  # should print 6, 9
```

8.5 Example: Pipelined Method for Finding Primes

A common example of parallel algorithms consists of finding prime numbers. The connection to data science is not strong—though examples exist in experimental design and cryptography—but the utter simplicity of the operation makes it an excellent way to introduce **Rmpi**. We will take a *pipelined* approach.

8.5.1 Algorithm

Our function **primepipe**, to be shown shortly, has three arguments:

- **n**: the function returns the vector of all primes between 2 and **n**, inclusive

- **divisors**: the function checks each potential prime for divisibility by the numbers in this vector

- **msgsize**: the size of messages from the manager to the first worker

Here are the details:

This is the classical Sieve of Eratosthenes. We make a list of the numbers from 2 to n, then "cross out" all multiples of 2, then all multiples of 3, then all multiples of 5, and so on. After the crossing-out by 2s, 3s and 5s, for instance, our list 2,3,4,5,6,7,8,9,10,11,12 will now be

2 3 ~~4~~ ~~5~~ ~~6~~ 7 ~~8~~ ~~9~~ ~~10~~ 11 ~~12~~ ...

In the end, the numbers that haven't gotten crossed out are the primes.

It's very much like a factory assembly line. The first station on the line crosses out by 2s, the next station crosses out by 3s in whatever remains in the first station's output, etc. So a good name for this kind of algorithm would be an *assembly line*, but it is traditional to call it a *pipeline.*

The vector **divisors** "primes the pump," as it were. We find a small set of primes using nonparallel means, and then use those in the larger parallel problem. But what range do we need for them? Reason as follows.

If a number i has a divisor k larger than $\sqrt{n}$, it must then have one (specifically, the number i/k) smaller than that value. Thus in crossing out all multiples of k, we need only consider values of k up to $\sqrt{n}$. So, in order to achieve our goal of finding all the primes up through n, we take our **divisors** vector to be all the primes up through $\sqrt{n}$.

The function **serprime()** in the code to be presented in Section 8.5.2 will do that. For example, say n is 1000. Then we first find all the primes less than or equal to $\sqrt{1000}$, using our nonparallel function, and use the result of that as input to the pipelined function, **primepipe()**, to find all the primes through 1000.

To do all this, I first got **Rmpi** running as in Section 8.4.1, and placed the code in Section 8.5.2 in a file **PrimePipe.R**. I then ran as follows:

```
> source("PrimePipe.R")
> dvs <- serprime(ceiling(sqrt(1000)))
> dvs
 [1]  2  3  5  7 11 13 17 19 23 29 31
> primepipe(1000,dvs,100)
  [1]   2   3   5   7  11  13  17  19  23  29  31  37  41  43  47  53  59  61
 [19]  67  71  73  79  83  89  97 101 103 107 109 113 127 131 137 139 149 151
 [37] 157 163 167 173 179 181 191 193 197 199 211 223 227 229 233 239 241 251
 [55] 257 263 269 271 277 281 283 293 307 311 313 317 331 337 347 349 353 359
 [73] 367 373 379 383 389 397 401 409 419 421 431 433 439 443 449 457 461 463
 [91] 467 479 487 491 499 503 509 521 523 541 547 557 563 569 571 577 587 593
[109] 599 601 607 613 617 619 631 641 643 647 653 659 661 673 677 683 691 701
[127] 709 719 727 733 739 743 751 757 761 769 773 787 797 809 811 821 823 827
[145] 829 839 853 857 859 863 877 881 883 887 907 911 919 929 937 941 947 953
[163] 967 971 977 983 991 997
```

Now, to understand the argument **msgsize**, consider the case n = 1000 above. Each worker will be responsible for its particular chunk of **divisors**. If we have two workers, then Process 0 (the manager) will "cross out" multiples of 2; Process 1 (the first worker) will handle multiples of 3, 5, 7, 11 and 13; Process 2 will handle k = 17, 19, 23, 29 and 31. So, Process 0 will cross out the multiples of 2, and send the remaining numbers, the odds, to Process 1. The latter will eliminate the multiples of 3, 5, 7, 11 and 13, and pass what is left to Process 2. What emerges from the latter is our final result, which Process 2 returns to Process 0.

The argument **msgsize** specifies the size of chunks of odds that process 0 sends to process 1. More on this point later.

8.5.2 The Code

```
# Rmpi code to find prime numbers

# for illustration purposes, not intended to be
# optimal, e.g. needs better load balancing

# returns vector of all primes in 2..n; the vector
# "divisors" is used as a basis for a Sieve of
# Erathosthenes operation; must have n <=
# (max(divisors)^2) and n even; "divisors" could
# first be found, for instance, by applying a serial
# prime-finding method on 2..sqrt(n), say
```

```
# with serprime() below

# the argument "msgsize" controls the chunk size in
# communication from the manager to the first worker,
# node 1

# manager code
primepipe <- function(n, divisors ,msgsize) {
   # supply the workers with the functions they need
   mpi.bcast.Robj2slave(dowork)
   mpi.bcast.Robj2slave(dosieve)
   # start workers, instructing them to each
   # run dowork();
   # note nonblocking call
   mpi.bcast.cmd(dowork,n, divisors ,msgsize)
   # remove the evens right away
   odds <- seq(from=3,to=n,by=2)
   nodd <- length(odds)
   # send odds to node 1, in chunks
   startmsg <- seq(from=1,to=nodd,by=msgsize)
   for (s in startmsg) {
      rng <- s:min(s+msgsize-1,nodd)
      # only one message type, 0
      mpi.send.Robj(odds[rng], tag=0, dest=1)
   }
   # send end-data sentinel, chosen to be NA
   mpi.send.Robj(NA,tag=0,dest=1)
   # wait for and receive results from last node,
   # and return the result;
   # don't forget the 2, the first prime
   # ID of last process
   lastnode <- mpi.comm.size()-1
   c(2,mpi.recv.Robj(tag=0,source=lastnode))
}

# worker code
dowork <- function(n, divisors ,msgsize) {
   me <- mpi.comm.rank()
   # which chunk of "divisors" is mine?
   lastnode <- mpi.comm.size()-1
   ld <- length(divisors)
   tmp <- floor(ld / lastnode)
   mystart <- (me-1) * tmp + 1
```

```
   myend <- mystart + tmp - 1
   if (me == lastnode) myend <- ld
   mydivs <- divisors[mystart:myend]
   # "out" will eventually contain the final results
   if (me == lastnode) out <- NULL
   # keep receiving a chunk from my predecessor node,
   # filtering it according to mydivs, sending what
   # survives to my successor node
   pred <- me - 1
   succ <- me + 1
   repeat {
      msg <- mpi.recv.Robj(tag=0, source=pred)
      # anything left to process?
      if (me < lastnode) {
         if (!is.na(msg[1])) {
            # do the crossouts for my range
            # of divisors
            sieveout <- dosieve(msg, mydivs)
            # send what remains to next node
            mpi.send.Robj(sieveout, tag=0, dest=succ)
         } else {
            # no more coming, so relay sentinel
            mpi.send.Robj(NA, tag=0, dest=succ)
            return()
         }
      } else { # I am the last node
         if (!is.na(msg[1])) {
            sieveout <- dosieve(msg,mydivs)
            out <- c(out,sieveout)
         } else {
            # no more coming, so send results
            # to manager
            mpi.send.Robj(out, tag=0, dest=0)
            return()
         }
      }
   }
}

# check divisibility of x by divs
dosieve <- function(x,divs) {
   for (d in divs) {
      x <- x[x %% d != 0 | x == d]
```

```
   }
   x
}

# serial prime finder; can be used to generate
# divisor list
serprime <- function(n) {
   nums <- 1:n
   # all in nums assumed prime until shown otherwise
   prime <- rep(1,n)
   maxdiv <- ceiling(sqrt(n))
   for (d in 2:maxdiv) {
      # don't bother dividing by nonprimes
      if (prime[d])
         # try divisor d on numbers not yet
         # found nonprime
         prime[prime !=0 & nums > d & nums %% d == 0]
            <- 0
   }
   nums[prime != 0 & nums >= 2]
}
```

8.5.3 Timing Example

Let's try the case n = 10000000. The serial code took time 424.592 seconds.

Let's try it in parallel on a network of PCs, for first two, then three and then four workers. with various values for **msgsize**. The results are shown in Table 8.1.

The parallel version was indeed faster than the serial one. This was partly due to parallelism and partly to the fact that the parallel version is more efficient, since the serial algorithm does more total crossouts. A fairer comparison might be a recursive version of **serprime()**, which would reduce the number of crossouts. But there are other important facets of the timing numbers.

First, as expected, using more workers produced more speed, at least in the range tried here. Note, though, that the speedup was not linear. The best time for three workers was only 30% better than that for two workers, compared to a "perfect" speedup of 50%. Using four workers instead of two yields only a 53% gain. Second, we see that **msgsize** is an important factor, explained in the next section.

msgsize	2 workers	3 workers	4 workers
1000	59.487	58.175	47.248
5000	22.855	17.541	15.454
10000	19.230	14.734	12.522
15000	19.198	14.874	12.689
25000	22.516	18.057	15.591
50000	23.029	18.573	16.114

Table 8.1: Timings, Prime Number Finding

8.5.4 Latency, Bandwdith and Parallelism

Another salient aspect here is that **msgsize** matters. Recall Section 2.5, especially Equation (2.1). Let's see how they affect things here.

In our timings above, setting the **msgsize** parameter to the lower value, 1000, results in having more chunks, thus more times that we incur the network latency. On the other hand, a value of 50000 yields less parallelism—there would be no parallelism at all with a chunk size of 10000000/2—and thus impedes are ability to engage in latency hiding (Section 2.5), in which we try to overlap computation and communication; this reduces speed.

8.5.5 Possible Improvements

There are a number of ways in which the code could be improved algorithmically. Notably, we have a serious load balance problem (Section 2.1). Here's why:

Suppose for simplicity that each process handles only one element of **divisors**. Process 0 then first removes all multiples of 2, leaving n/2 numbers. Process 1 then removes all multiples of 3 from the latter, leaving n/3 numbers. It can be seen from this that Process 2 has much less work to do that Process 1, and Process 3 has a much lighter load than process 2, etc.

One possible solution might be to have the code do its partitioning of the vector **divisors** in an uneven way, assigning larger chunks of the vector to the later processes.

Note that the code sends data messages via the functions **mpi.send.Robj()**

and **mpi.recv.Robj()**, rather than **mpi.send()** and **mpi.recv()**. The latter two would be more efficient, as the former two perform serialize/unserialize operations (Section 2.10), thus slowing things down, and are also slower in terms of memory allocation (Section 8.6). Nevertheless, **Rmpi** is such a rich, complex package that it is best to introduce it in a simple manner, hence our use of the somewhat slower functions.

8.5.6 Analysis of the Code

So, let's look at the code. First, a bit of housekeeping. Just as with **snow**, we need to send the workers the functions they'll use:

```
mpi.bcast.Robj2slave(dowork)
mpi.bcast.Robj2slave(dosieve)
```

"Robj" of course stands for "R object," so we can send anything, in this case sending functions. This is an illustration of a nice feature of **Rmpi** over MPI.

Next, we get the ball rolling, by having the manager send data to the first worker:

```
odds <- seq(from=3,to=n,by=2)
nodd <- length(odds)
startmsg <- seq(from=1,to=nodd,by=msgsize)
for (s in startmsg) {
   rng <- s:min(s+msgsize-1,nodd)
   mpi.send.Robj(odds[rng],tag=0,dest=1)
}
mpi.send.Robj(NA,tag=0,dest=1)
```

Remember, our prime finding algorithm consists of first eliminating multiples of 2, then of 3 and so on. Here the manager takes that first step, eliminating the even numbers.

Note the fact that the **for** loop implements our plan for the manager to send out the odd numbers in chunks, rather than all at once. This is crucial to parallelization. If we don't use the advanced (and difficult) technique of nonblocking I/O (Section 8.7.1), then sending out the entire vector **odds**, and acting similarly at the workers, would give us no parallelism at all; we would have only one worker doing "crossing out" at a time. By sending data in chunks, we can keep everyone busy at once, as soon as the pipeline fills.

As noted earlier, the parameter **msgsize** controls the tradeoff between the computation/communication overlap, and the overhead of launching a message. A larger value means fewer times we pay the latency price, but less parallelism.

Note that each worker needs to know when there will be no further input from its predecessor. Sending an NA value serves that purpose.

In the call to **mpi.send.Robj()** above,

```
mpi.send.Robj(odds[rng],tag=0,dest=1)
```

the argument **tag=0** means we are considering this message to be of type 0. Message types are programmer-defined, and we only have one kind of message here. But in some applications we might define several different types. Even in our code here, we could define a second type, to denote a "no more data" condition, instead of signifying the condition by sending an NA value, as we do here. MPI gives the receiver the ability to receive the next message of a specified type, or to receive any message without regard to type and then ask MPI what type it is. Here we could have defined type 1 to mean "no more data," and after executing a receive, we would check for type 1.

The argument **dest=1** means, "send this message to Process 1," i.e., the first worker. Since MPI numbers processes starting from 0 rather than 1, **Rmpi** does the same, with the manager being Process 0.

Next the manager starts up the workers. Technically, they have already been running. For example, we may have made an earlier call to the **Rmpi** function **mpi.spawn.Rslaves()**. But they are not doing any useful work yet. Each worker, upon startup, enters a loop in which it repeatedly executes **mpi.bcast.cmd()**, with "bcast" standing for "broadcast." There are some subtle issues here.

The name of the function **mpi.bcast.cmd()** is a little confusing, because it sounds like all the processes are broadcasting, but it means they are all *participating* in a broadcast operation. Here the manager is doing the broadcast, and the workers are receiving that broadcast.

So, consider what happens when the manager executes

```
mpi.bcast.cmd(dowork,n,divisors,msgsize)
```

As noted, each worker had executed a call to **mpi.bcast.cmd()**, which was hanging, waiting for the manager to make such a call. When the manager does so in the above code, each worker's call to **mpi.bcast.cmd()**

will complete, by executing the request from the manager, in this case a command to run the **dowork()** function.

The command broadcast by the manager here tells the workers to execute **dowork(n,divisors,msgsize)**. They will thus now be doing useful work, in the sense that they are now running the application, though they still must wait to receive their data.

Eventually the last worker will send the final list of primes back to the manager, which will receive it, and return the result to the caller:

```
lastnode <- mpi.comm.size()-1
c(2,mpi.recv.Robj(tag=0,source=lastnode))
```

Note that since the manager had removed the multiples of 2 originally, the number 2 won't be in what is received here. Yet of course 2 is indeed a prime, so we need to add it to the list.

The function **mpi.comm.size()** returns the *communicator* size, the total number of processes, including the manager. Recalling that the latter is Process 0, we see that the last worker's process number will be the communicator size minus 1. In more advanced MPI applications, we can set up several communicators, i.e., several groups of processes, rather than just one, our case here. A broadcast then means transmitting a message to all processes in that communicator.

So, what about **dowork()**, the function executed by the workers? First, note that worker i must receive data from worker i-1 and send data to worker i+1. Thus the worker needs to know its process ID number, or *rank* in MPI parlance:

```
me <- mpi.comm.rank()
```

Now the worker must decide which of the divisors it will be responsible for. This will be a standard chunking operation:

```
ld <- length(divisors)
tmp <- floor(ld / lastnode)
mystart <- (me-1) * tmp + 1
myend <- mystart + tmp - 1
if (me == lastnode) myend <- ld
mydivs <- divisors[mystart:myend]
```

An alternative would have been to use **mpi.scatter()**, distributing the vector **divisors** via a scatter operation. Since the **divisors** vector will be short, this approach wouldn't give us a performance boost, but it would make the code a bit more elegant.

The heart of **dowork()** is a large **repeat** loop, in which the worker repeatedly receives data from its predecessor,

```
msg <- mpi.recv.Robj(tag=0,source=pred)
```

does the necessary "crossing out,"

```
sieveout <- dosieve(msg,mydivs)
```

and sends the result to its successor worker,

```
mpi.send.Robj(sieveout,tag=0,dest=succ)
```

In the case of the final worker, it accumulates its "crossing out" results in a vector **out**,

```
sieveout <- dosieve(msg,mydivs)
out <- c(out,sieveout)
```

which it sends to the manager, Process 0, at the end:

```
mpi.send.Robj(out,tag=0,dest=0)
```

The "crossing out" function, **dosieve()** is straightforward, but note that we do try to make good use of vectorization:

```
x <- x[x %% d != 0 | x == d]
```

8.6 Memory Allocation Issues

Memory allocation is a major issue, both in this application and many others, thus worth spending some extra time on here. The problem is that when a message arrives at a process, **Rmpi** needs to have a place to put it.

Recall that although our prime-finding code above called **mpi.recv.Robj()**, receiving a general R object, the more basic receive operation is **mpi.recv()**. If we call the latter, we must set up a buffer for it, e.g. **b** in

```
b <- double(100000)
b <- mpi.recv(x=b,type=2,source=5)
```

If the receive call is within a loop, the overhead of repeatedly setting up buffer space may be substantial. This of course would be remedied by moving the statement

```
b <- double(100000)
```

to a position preceding the loop.[4]

With **mpi.recv.Robj()**, this memory allocation overhead occurs "invisibly." If the function is called from within a loop, there is potentially a reallocation at every iteration. So, while this type of receive call is more convenient, you should not be fooled into thinking there are no memory issues. Thus we may attain better efficiency from **mpi.recv()** than from **mpi.recv.Robj()**. As mentioned, the latter also suffers some slowdown from serialization.

On the other hand, if we use **mpi.recv()** and set the memory allocation before the loop, we must allocate enough memory for the largest message that might be received. This may be wasteful of memory, and if memory space is an issue, this is a problem that must be considered.

8.7 Message-Passing Performance Subtleties

In message-passing systems, even innocuous-looking operations can have lots of important subtleties. This section will present an overview.

8.7.1 Blocking vs. Nonblocking I/O

The call

```
mpi.send(x,type=2,tag=0,dest=8)
```

sends the data in **x**. But when does the call return? The answer depends on the underlying MPI implementation. In some implementations, probably most, the call returns as soon as the space **x** is reusable, as follows. **Rmpi** will call MPI, which in turn will call network-send functions in the operating system. That last step will involve copying the contents of **x** from one's R

[4]Even this may not be enough. R has a *copy on write* policy, meaning that if a vector element is changed, the memory for the vector may be reallocated. The word "may" is key here, and recent versions of R attempt to reduce the number of reallocations, but there is never any guarantee on this.

program to space in the OS, after which **x** is reusable. The point is that the call could return long before the data reaches the receiver.

Other implementations of MPI, though, wait until the destination process, number 8 in the example above, has received the transmitted data. The call to **mpi.send()** at the source process won't return until this happens.

Due to network delays, there could be a large performance difference between the two MPI implementations. There are also possible implications for deadlock (Section 8.7.2).

In fact, even with the first kind of implementation, there may be some delay. For such reasons, MPI offers *nonblocking* send and receive functions, for which **Rmpi** provides the interfaces such as **mpi.isend()** and **mpi.irecv()**. This way you can have your code get a send or receive started, do some other useful work, and then later check back to see if the action has been completed, using a function such as **mpi.test()**.

8.7.2 The Dreaded Deadlock Problem

Consider code in which Processes 3 and 8 trade data:

```
me <- mpi.comm.rank()
if (me == 3) {
   mpi.send(x,type=2,tag=0,dest=8)
   mpi.recv(y,type=2,tag=0,source=8)
} else if (me == 8){
   mpi.send(x,type=2,tag=0,dest=3)
   mpi.recv(y,type=2,tag=0,source=3)
}
```

If the MPI implementation has send operations block until the matching receive is posted, then this would create a *deadlock* problem, meaning that two processes are stuck, waiting for each other. Here Process 3 would start the send, but then wait for an acknowledgment from 8, while 8 would do the same and wait for 3. They would wait forever.

This arises in various other ways as well. Suppose we have the manager launch the workers via the call

```
mpi.bcast.cmd(dowork,n,divisors,msgsize)
```

This sends the command to the workers, then immediately returns. By contrast,

```
res <- mpi.remote.exec(dowork,n,divisors,msgsize)
```

would make the same call at the workers, but would wait until the workers were done with their work before returning (and then assigning the results to **res**). Now suppose the function **dowork()** does a receive from the manager, and suppose we use that second call above, with **mpi.remote.exec()**, after which the manager does a send operation, intended to be paired with the receive ops at the workers. In this setting, we would have deadlock.

Deadlock can arise in shared-memory programming as well (Chapter 4), but the message-passing paradigm is especially susceptible to it. One must constantly beware of the possibility when writing message-passing code.

So, what are the solutions? In the example involving Processes 3 and 8 above, one could simply switch the ordering:

```
me <- mpi.comm.rank()
if (me == 3) {
   mpi.send(x,type=2,tag=0,dest=8)
   mpi.recv(y,type=2,tag=0,source=8)
} else if (me == 8){
   mpi.recv(y,type=2,tag=0,source=3)
   mpi.send(x,type=2,tag=0,dest=3)
}
```

MPI also has a combined send-receive operation, interfaced to from **Rmpi** via **mpi.sendrecv()**.

Another way out of deadlock is to use the nonblocking sends and/or receives, at the cost of additional code complexity.

8.8 Further Reading

MPI includes a large variety of functions, some of which involve delicate asynchronous operations. Several good guides are available, including *Parallel Programming with MPI*, by Peter Pacheco, Morgan Kaufman, 1997.

Chapter 9

MapReduce Computation

As the world emerged into an era of Big Data, demand grew for a computing paradigm that (a) is generally applicable and (b) works on *distributed* data. The latter term means that data is physically distributed over many chunks, possibly on different disks and maybe even different geographical locations. Having the data stored in a distributed manner facilitates parallel computation — different chunks can be read simultaneously — and also enables us to work with data sets that are too large to fit into the memory of a single machine. Demand for such computational capability led to the development of various systems using the *MapReduce* paradigm.

MapReduce is really a form of the scatter-gather pattern we've seen frequently in this book, with the added feature of a sorting operation added in the middle. In rough form, it works like this. The input is in a (distributed) file, fed into the following process:

- **Map phase:** There are various parallel processes known as *mappers*, all running the same code. For each line of the input file, the mapper handling that chunk of the file reads the line, processes it in some way, and then emits an output line, consisting of a key-value pair.
- **Shuffle/sort phase:** All the mapper output lines that share the same key are gathered together.
- **Reduce phase:** There are various parallel processes known as *reducers*, all running the same code. Each reducer will work on its own set of keys, i.e., for any given key, all mapper output lines having the same key will go to the same reducer. Moreover, the lines fed into any given reducer will be sorted by key.

9.1 Apache Hadoop

At this writing, the most popular MapReduce software package is Hadoop. It is written in Java, and is most efficiently used in that language (or C++), but it includes a *streaming* feature that enables one to use Hadoop from essentially any language, including R. Hadoop includes its own distributed file system, unsurprisingly called the Hadoop Distributed File System (HDFS). Note that one advantage of HDFS is that it is replicated, thus achieving some degree of fault tolerance.

9.1.1 Hadoop Streaming

As noted, Hadoop is really written for Java or C++ applications. However, Hadoop can work with programs in any language under Hadoop's streaming option, by reading from STDIN and writing to STDOUT, in text, line-oriented form in both cases.

Input to the mappers is from an HDFS file, and output from the reducers is again to an HDFS file, one chunk per reducer. The file line format is

```
key \t data
```

where `\t` is the Tab character.

The usage of text format does cause some slowdown in numeric programs, for the conversion of strings to numbers and vice versa, but again, Hadoop is not designed for maximum efficiency.

9.1.2 Example: Word Count

The typical "Hello World," introductory example is word count for a text file. The mapper program breaks a line into words, and emits (key,value) pairs in the form of (word,1). (If a word appears several times in a line, there would be several pairs emitted for that word.) In the Reduce stage, all those 1s for a given word are summed, giving a frequency count for that word. In this way, we get counts for all words.

Here is the mapper code:

```
#! /usr/bin/env Rscript

# wordmapper.R

si <- file("stdin",open="r")
while (length(inln <-
      scan(si,what="",nlines=1,quiet=TRUE,
         blank.lines.skip=FALSE))) {
   for (w in inln) cat(w,"\t 1\n")
}
```

And here is the reducer:

```
#! /usr/bin/env Rscript

# wordreducer.R

si <- file("stdin",open="r")
oldword <- ""

while (length(inln <- scan(si,what="",nlines=1,
      quiet=TRUE))) {
   word <- inln[1]
   if (word != oldword) {
      if (oldword != "")
         cat(oldword,"\t ",count,"\n")
      oldword <- word
      count <- 1
   } else {
      count <- count + as.integer(inln[2])
   }
}
```

The above code is not very refined, for instance treating *The* as different from *the*. The main point, though, is just to illustrate the principles.

9.1.3 Running the Code

I ran the following code from the top level of the Hadoop directory tree, with obvious modifications possible for other run points. First, I needed to place my data file, **rnyt**,[1]

```
$ bin/hadoop fs -put ../rnyt rnyt
$ bin/hadoop jar contrib/streaming/*.jar \
     -input rnyt \
     -output wordcountsnyt \
     -mapper ../wordmapper.R
     -reducer ../wordreducer.R
```

That first command specifies that it will be for the file system (**fs**), and that I am placing a file in that system. My ordinary version of the file was in my home directory.

Hadoop, being Java-based, runs Java archive, **.jar** files. So, the second command above specifies that I want to run in streaming mode. It also states that I want the output to go to a file **wordcountsnyt** in my HDFS system. Finally, I specify my mapper and reducer code files in my HDFS system, after which I ran the program. I allowed Hadoop to use its default values for the number of mappers and reducers, and could have specified them above if desired.

Recall that the final output comes in chunks in the HDFS. Here's how to check (some material not shown), and to view the actual file contents:

```
$ bin/hadoop fs -ls wordcountsnyt

Found 3 items
-rw-r--r--   1 ... /user/matloff/wordcountsnyt/_SUCCESS
drwxr-xr-x   - ... /user/matloff/wordcountsnyt/_logs
-rw-r--r--   1 ... /user/matloff/wordcountsnyt/part-00000
$ bin/hadoop fs -cat wordcountsnyt1/part-00000
1          NA
$2         1
1,600      1
18th       1
1991,      1
1996,      1
2009       1
2009,      1
250,000            1
6,         1
7,         1
```

[1]The file was the contents of the article "Data Analysts Captivated by Rs Power," *New York Times*, January 6, 2009.

```
A       1
ASHLEE  1
According       1
America ,       1
Analysts        2
Anne    1
Another         1
Apache ,        1
Are     1
At      1
...
```

So, in HDFS, one distributed file is stored as a directory, with the chunks in **part-00000**, **part-00001** and so on. We only had enough data to fill one chunk here.

9.1.4 Analysis of the Code

The first thing to notice is that these two R files are not executed directly by R, but instead under Rscript. This is standard for running R in *batch*, i.e., noninteractive, mode.

Next, as noted earlier, input to the mappers is from STDIN, in this case from the redirected file **rnyt** in my HDFS, seen here in the identifier **si** in the mapper. Final output is to STDOUT, redirected to the specified HDFS file, hence the call to **cat()** in the reducer. The mapper output goes to the shuffle and then to the reducers, again using STDOUT, visible here in the call to **cat()** in the mapper code.

The reader might be wondering here about the line

```
count <- count + as.integer(inln[2])
```

in the reducer. The way things have been described so far, it would seem that the expression **as.integer(inln[2])** — a word count output from a mapper — should always be 1. However, there is more to the story, as Hadoop also allows one to specify *combiner* code, as follows.

Remember, all the communication, e.g. from the mappers to the shuffler, is via our network, with one (key,value) pair per network message. So, we may have an enormous number of short messages, thus incurring the network latency penalty many times, as well as huge network congestion due to, for instance, many mappers trying to use the network at once. The solution is to have each mapper try to coalesce its messages before sending to the shuffler.

The coalescing is done by a *combiner* specified by the user. Often the combiner will be the same as the reducer. So, what occurs is that each mapper will run the reducer on its own mapper output, then send the combiner output to the shuffler, after which it goes to the reducers as usual.

Thus in our word count example here, when a line arrives at a reducer, its count field may already have a value greater than 1. The combiner code, by the way, is specified via the **-combiner** field in the run command, like **-mapper** and **-reducer**.

9.1.5 Role of Disk Files

As noted, Hadoop has its own file system, HDFS, which is built on top of the native OS' file system of the machines. It is replicated for the sake of reliability, with each HDFS block existing in at least 3 copies, i.e., on at least 3 separate disks. Very large files are possible, in some cases spanning more than one disk/machine.

Disk files play a major role in Hadoop programs:

- Input is from a file in the HDFS system.
- The output of the mappers goes to temporary files in the native OS' file system.
- Final output is to a file in the HDFS system. As noted earlier, that file may be distributed across several disks/machines.

Note that by having the input and output files in HDFS, we minimize communications costs in shipping the data between nodes of a cluster. The slogan used is "Moving computation is cheaper than moving data." However, all that disk activity can be quite costly in terms of run time.

9.2 Other MapReduce Systems

As of late 2014, there has been increasing concern regarding Hadoop's performance. One of the problems is that one cannot keep intermediate results in memory between Hadoop runs. This is a serious problem, for instance, with iterative or even multi-pass algorithms.

The Spark package now being developed aims to remedy many of Hadoop's shortcomings. Early reports indicate some drastic speed improvements, while retaining the ability to read HDFS files, and continuing to have fault tolerance features.

9.3 R Interfaces to MapReduce Systems

Given the widespread usage of Hadoop, a number of R interfaces have been developed. The most popular is probably **rmr**, developed by Revolution Analytics, and RHIPE, written by Saptarshi Guha as part of his PhD dissertation. An R interface package, **sparkr**, is available.

9.4 An Alternative: "Snowdoop"

So, what does Hadoop really give us? The two main features are (a) distributed data access and (b) an efficient distributed file sort. Hadoop works well for many applications, but a realization developed that Hadoop can be very slow, and very limited in available data operations.

Both of those shortcomings are addressed to a large extent by the new kid on the block, Spark. Spark is apparently much faster than Hadoop, sometimes dramatically so, due to strong caching ability and a wider variety of available operations. Recently **distributedR** has also been released, again with the goal of using R on voluminous data sets, and there is also the more established **pbdR**.

But even Spark suffers a very practical problem, shared by the others mentioned above. All of these systems are complicated. There is a considerable amount of configuration to do, worsened by dependence on infrastructure software such as Java or MPI, and in some cases by interface software such as rJava. Some of this requires systems knowledge that many R users may lack. And once they do get these systems set up, they may be required to design algorithms with world views quite different from R, even though they are coding in R.

So, do we really need all that complicated machinery? Hadoop and Spark provide efficient distributed sort operations, but if one's application does not depend on sorting, we have a cost-benefit issue here.

Here is an alternative, which I call "Snowdoop": One simply does one's own chunking of files into distributed mini-files, and then uses Snow or some other general R tool on those files.

9.4.1 Example: Snowdoop Word Count

Let's use as our example word count, the "Hello World" of MapReduce. As noted earlier, word count determines which words are in a text file, and calculates frequency counts for each distinct word:

```
# each node executes this function
wordcensus <- function(basename,ndigs) {
   fname <- filechunkname(basename,ndigs)
   words <- scan(fname,what="")
   tapply(words,words,length, simplify=FALSE)
}

# manager
fullwordcount <- function(cls,basename,ndigs) {
   setclsinfo(cls)  # give workers ID numbers, etc.
   counts <-
      clusterCall(cls,wordcensus,basename,ndigs)
   addlistssum <- function(lst1,lst2)
      addlists(lst1,lst2,sum)
   Reduce(addlistssum,counts)
}
```

The above code makes use of the following routines, which are general and are used in many "Snowdoop" applications. These and other Snowdoop utilities are included in my **partools** package (Section 3.5). Here are the call forms:

```
# give each cluster node an ID, in a global
# partoolsenv$myid; number of workers in
# partoolsenv$ncls; at each worker,
# load partools and set the R search path to
# that of the manager
setclsinfo(cls)

# "add" R lists lst1, lst2, applying the operation
# 'add' to elements in common, copying non-null others
addlists(lst1,lst2,add)
```

```
# form the file name basename.i, where i is the ID
# of this node unless nodenum is specified
filechunkname(basename,ndigs,nodenum=NULL)
```

All pure R! No Java, no configuration. Indeed, it's worthwhile comparing to the word count example in the **sparkr** distribution. There we see calls to **sparkr** functions such as **flatMap()**, **reduceByKey()** and **collect()**. But the **reduceByKey()** function is pretty much the same as R's tried and true **tapply()**. The **collect()** function is more or less our Snowdoop library function **addlists()**. So, again, there is no need to resort to Spark, Hadoop, Java and so on; we just use ordinary R.

We are achieving the parallel-read advantage of Hadoop and Spark,[2] *while avoiding the Hadoop/Spark configuration headaches and while staying with the familiar R programming paradigm.* In many cases, this should be a highly beneficial tradeoff for us.

Of course, this approach lacks the fault tolerance feature of Hadhoop and Spark can, which can be quite advantageous. And as of this writing, it is not yet clear how well this scales, e.g. how well the **parallel** package works with very large numbers of nodes. But Snowdoop is an attractive approach for many applications.

9.4.2 Example: Snowdoop k-Means Clustering

Let's look again at the k-means clustering example of Section 4.9.

```
# k-means clustering, using Snowdoop

# arguments:
#
#    xname: name of chunked data (typical read in
#           earlier from chunked file
#    nitrs: number of  iterations to perform
#    nclus:  number of  clusters to form
#    ctrs:  the matrix of initial centroids

# assumes:
#
#    setclsinfo already called
#    a cluster will never become empty
```

[2]Note that neither Hadoop, Spark nor Snowdoop will achieve full parallel reading if the file chunks are all on the same disk.

```
kmeans <- function(cls,xname,nitrs,ctrs) {
   addlistssum <-
      function(lst1,lst2) addlists(lst1,lst2,sum)
   for (i in 1:nitrs) {
      # for each data point, find the nearest centroid,
      # and tabulate; at each worker and for each
      # centroid, we compute a vector whose first
      # component is the count of the number of
      # data points whose nearest centroid is that
      # centroid, and whose remainining portion
      # is the sum of all such data points
      tmp <- clusterCall(cls,findnrst,xname,ctrs)
      # sum over all workers
      tmp <- Reduce(addlistssum,tmp)
      # compute new centroids
      for (i in 1:nrow(ctrs)) {
         tmp1 <- tmp[[as.character(i)]]
         ctrs[i,] <- (1/tmp1[1]) * tmp1[-1]
      }
   }
   ctrs
}

findnrst <- function(xname,ctrs) {
   require(pdist)
   x <- get(xname)
   dsts <- matrix(pdist(x,ctrs)@dist,ncol=nrow(x))
   # dsts[,i] now has the distances from row i of x
   # to the centroids
   nrst <- apply(dsts,2,which.min)
   # nrst[i] tells us the index of the centroid
   # closest to row i of x
   mysum <- function(idxs,myx)
      c(length(idxs),colSums(x[idxs,,drop=FALSE]))
   tmp <- tapply(1:nrow(x),nrst,mysum,x)
}

test <- function(cls) {
   m <- matrix(c(4,1,4,6,3,2,6,6),ncol=2)
   formrowchunks(cls,m,"m")
   initc <- rbind(c(2,2),c(3,5))
   kmeans(cls,"m",1,initc) }
```

So again, we are using chunked files as in Hadoop, but writing ordinary R code, e.g. **tapply()** and **Reduce()**. But most important, the data at each worker persists across iterations. In Hadoop, it would be reread from disk at each iteration, and in Spark, we'd need to request caching, but here it comes for free, no special effort needed.

9.5 Further Reading

Parallel R, by Q. Ethan McCallum and Stephen Weston, O'Reilly 2011, has quite a bit on using R with Hadoop. The vignette included with the **partools** package contains many examples of use of that package,

Chapter 10

Parallel Sorting and Merging

One of the most common types of computation is sorting, the main subject of this chapter. A related topic is *merging*, meaning to combine two sorted vectors into one large sorted vector.

Sorting is not an embarrassingly parallel operation (Section 2.11). Accordingly, many different types of parallel sorts have been invented. We'll cover some introductory material here.

10.1 The Elusive Goal of Optimality

There is a vast literature on sorting methods, including for parallel computing. But the "best" one depends somewhat on the nature of the data, thus on the nature of the application, and even more on the nature of the computing platform. What works well on a multicore machine, for instance, may do poorly on a GPU.

Entire books have been written on this topic. Here, though, we briefly discuss some general approaches, and then note some excellent libraries to draw upon. In particular, the Thrust and **Rth** libraries (Chapter 7) will play a prominent role.

10.2 Sorting Algorithms

There is no such thing as a free lunch—famous term from economics

As mentioned, there is a tremendous variety of sorting algorithms. In this section we'll discuss a few common ones, to give the reader an idea of how these work. Unfortunately, each of them has its set of drawbacks, especially when implemented in parallel.

10.2.1 Compare-and-Exchange Operations

Many sorting algorithms make heavy use of *compare-and-exchange operations*. In the case of two numbers, x and y, this means

```
if (x > y) then swap x and y
```

Due to overhead issues, such as a need to amortize network latency (Section 2.5) or to reduce cache coherency actions (Section 2.5.1.1), this is often done in group form, i.e., the compare-and-swap operation works on two groups of numbers. The definition of "<" in the group case differs from one algorithm to another, as will be seen below.

10.2.2 Some "Representative" Sorting Algorithms

Each of these algorithms can be described simply, and outlines of them will be shown here; details may be found on a myriad of websites, including specific implementations in C or other languages.

Here we assume we have an R vector **x** of length n to be sorted. For simplicity, assume the n values are distinct.

- **The Much-Maligned Bubble Sort**

  This algorithm is widely taught as an example of what *not* to do. To sort n numbers on a serial platform, it takes $O(n^2)$ time, compared to the $O(n \log n)$ time for better algorithms. However, as we will see shortly, in a parallel context the bubble sort can be useful.

  The algorithm is quite simple. We start with **x[1]**, performing a compare-and-exchange operation with **x[2]**. We then do compare-and-exchange between **x[2]** and **x[3]** (remember, **x[2]** may now be the original **x[1]**), and keep going, finally performing compare-and-exchange at **x[n-1]**. This will be a total of *n-1* steps.

We then go back to **x[2]**, moving rightward through the whole vector in the same manner. This will involve *n-2* compare-and-exchange operations. We then do this at **x[3]**, with *n-3* operations, and so on.

So, we have n-1 passes through the vector, with the number of compare-and-exchange operations being

$$n - 1 + n - 2 + n - 3 + ... + 1$$

which by a famous mathematical formula boils down to (n-1)n/2, i.e., $O(n^2)$.

A variant is Even/Odd Transposition Sort. In the first step, all elements of **x** at even-numbered positions do compare-and-exchange with their right-hand neighbors; for instance, **x[8]** will be compared-and-exchanged with its right-hand neighbor **x[9]**. At the second step, the operation is with the left-hand neighbors. In the third step, the exchange is back to the right-hand direction, and so on.

This proceeds for approximately n steps, and again, at every step we perform about *n/2* comparisons. So, we have a time complexity of $O(n^2)$, as with the Bubble Sort.

If we wish to chunk the data, we do compare-and-exchange using neighboring chunks, defining that operation in various ways. For example, one can have each thread in a pair send the other all its data, and then have the lower-ID thread retain the lower half of the result, while the other thread keeps the upper half.

The algorithm thus can easily be parallelized. In general it is not the algorithm of choice, but on some special-purpose hardware forms it can work very well.

- **Quicksort**

  Start with **x[1]** as our *pivot*, i.e., our basis for comparison. Determine which elements of **x** are less than **x[1]**, and call them the "small group." Say there are **k** of them. The others, i.e., the ones larger than **x[1]**, form the "large group." Write the small group back into **x**, writing the first at element 1 and so on; write the original **x[1]** at element **k+1**, and write the large group at elements **k+2** through **n**.

  Now apply the same manipulation twice, once to to the small group, and once to the large group. In the former case, take the new **x[1]** as the pivot, while in the latter case take the new **x[k+2]**.

  Keep going in this manner. Each time you form a group, split it into its own small and large groups, and rearrange within those two groups. In the end, the vector will be sorted in-place.

The following rough analysis shows the time complexity. We are dividing the data in halves (albeit not exactly half), then in fourths, then in eighths and so on. Eventually our groups are all of size 1, so the process takes approximately $\log_2 n$ steps. In each step, we compare all elements to their current pivots, thus about n operations. In other words, the time complexity is $O(n \log n)$.

Readers who are familiar with *recursive* programming will recognize the algorithm here. An outline would be

```
qs <- function(x) {
   if (length(x) <= 1) return()
   find small group
   find large group
   move small group to left part of x
   move large group to right part of x
   move x[1] to position k+1 within x
   x[small group] <- qs(x[small group])
   x[large group] <- qs(x[large group])
}
```

The recursive nature of the algorithm is seen in the fact that the function **qs()** calls itself! At first, that might seem like magic, but it really makes sense when one thinks about the way this is handled internally. Interested readers may wish to review from Section 4.1.2, and ponder how recursion is implemented internally. Note that recursion actually is not very efficient in R, though it can be used well in C/C++. Full C code will be presented in Section 10.4.

Quicksort is optimal in principle, but even a serial implementation must be done very carefully to achieve good efficinency. In parallel, all the horrors discussed in Chapter 2—memory contention, cache coherency, network overhead and so on—arise with a vengeance.

A variant, Hyperquicksort, was developed for *hypercube* network topologies, but is applicable generally, especially for distributed data. It is discussed in Section 10.7.1.

- **Mergesort**

  Say n is 100000 and we have four threads. We could assign **x[1]** through **x[250000]** to the first thread, **x[250001]** through **x[500000]** to the second thread, and so on. Each thread then does a local sort of its chunk, and then all the sorted chunks are merged.

  So far, this is embarrassingly parallel. But the merge phase isn't. The latter is typically done in a tree-like manner. In our four-thread example above, say, we could have threads 1 and 2 merge their respective chunks, giving the result to thread 1, and at the same time have threads 3 and 4 do the same, giving this result to thread 3. Then threads 1 and 3 would merge their new chunks, and the vector would now be sorted.

In this manner, the merge phase would take $O(\log_2 n)$ steps, since at each step we are halving the number of active threads. At each step, all n elements of the vector get touched in some way, so as with Quicksort, the total time complexity is $O(n \log n)$. By the way, Thrust does include a merge function, **thrust::merge(x,y,z)**, that merges sorted vectors **x** and **y**, and places the result in **z**.

The major drawback of mergesorts is that many threads become idle at various points in the merge phase, thus robbing the process of parallelism.

- **Bucket Sort**

  This algorithm, sometimes called *sample sort*, is like a one-level Quicksort. Say we have three threads and 90000 numbers to sort. We could first take a small subsample, say of size 1000, and then (if we are in R) call **quantile()** to determine where the 0.33 and 0.67 quantiles are, which I'll call b and c. Thread 1 then handles all the numbers less than b, thread 2 handles the ones between b and c, and thread 3 handles the ones bigger than c. Each sorts its own group locally, then places the result back in the proper place in the vector. If for instance, thread 1 has 29532 numbers to process, it places the result of its sort in **x[1]** through **x[29532]**, and so on.

10.3 Example: Bucket Sort in R

Before resorting to C/C++, let's look at a pure R example. Here we implement Bucket Sort, using the **multicore** portion of the **parallel** package.

```
# mc.cores is the number of cores to use in computation
mcbsort <- function(x,ncores,nsamp=1000) {
   require(parallel)
   # get subsample to determine approximate quantiles
   samp <-
      sort(x[sample(1:length(x),nsamp,replace=TRUE)])
   # each thread will run dowork()
   dowork <- function(me) {
      # which numbers will this thread process?
```

```
      # (could also use quantile() here)
      k <- floor(nsamp / ncores)
      if (me > 1) mylo <- samp[(me-1) * k + 1]
      if (me < ncores) myhi <- samp[me * k]
      if (me == 1) myx <- x[x <= myhi] else
         if (me == ncores) myx <- x[x > mylo] else
            myx <- x[ x > mylo & x <= myhi]
      # this thread now sorts its assigned group
      sort(myx)
   }
   res <- mclapply(1:ncores,dowork,mc.cores=ncores)
   # string the results together
   c(unlist(res))
}

test <- function(n,ncores) {
   x <- runif(n)
   mcbsort(x,ncores=ncores,nsamp=1000)
}
```

This is a straightforward implementation of Bucket Sort; see the comments for details.

However, one should feel intuitively that this is not the best we can do. We'll do some timings shortly.

10.4 Example: Quicksort in OpenMP

```
// OpenMP example program:
// quicksort; not necessarily efficient

// exchange the elements pointed to by yi and yj
void swap(int *yi, int *yj)
{  int tmp = *yi;
   *yi = *yj;
   *yj = tmp;
}

// consider the section of x from x[low] to x[high],
// comparing each element to the pivot, x[low]; keep
// shuffling this section of x until, for some m,
// all the elements to the left of x[m] are <= the,
```

```
// pivot, and all the ones to the right
// are >= the pivot
int *separate(int *x, int low, int high)
{  int i,pivot,m;
   pivot = x[low];
   swap(x+low,x+high);
   m = low;
   for (i = low; i < high; i++) {
      if (x[i] <= pivot) {
         swap(x+m,x+i);
         m += 1;
      }
   }
   swap(x+m,x+high);
   return m;
}

// quicksort of the array z, elements zstart through
// zend; set the latter to 0 and n-1 in first call,
// where n is the length of z; firstcall is 1 or 0,
// according to whether this is the first of the
// recursive calls
void qs(int *z, int zstart, int zend, int firstcall)
{
   #pragma omp parallel
   {  int part;
      if (firstcall == 1) {
         #pragma omp single nowait
         qs(z,0,zend,0);
      } else {
         if (zstart < zend) {
            part = separate(z,zstart,zend);
            #pragma omp task
            qs(z,zstart,part-1,0);
            #pragma omp task
            qs(z,part+1,zend,0);
         }

      }
   }
}

// test code
```

```
main(int argc, char**argv)
{  int i,n,*w;
   n = atoi(argv[1]);
   w = malloc(n*sizeof(int));
   for (i = 0; i < n; i++) w[i] = rand();
   qs(w,0,n-1,1);
   if (n < 25)
      for (i = 0; i < n; i++) printf("%d\n",w[i]);
}
```

The code

```
if (firstcall == 1) {
   #pragma omp single nowait
   qs(z,0,zend,0);
```

gets things going. We want only one thread to execute the root of the recursion tree, hence the need for the **single** clause. The other threads will have nothing to do this round, but the root call sets up two new calls, each of which will again encounter the omp parallel pragma and the code

```
part = separate(z,zstart,zend);
#pragma omp task
qs(z,zstart,part-1,0);
```

Here the **task** directive states, "OMP system, please make sure that this subtree is handled by some thread eventually." If there are idle threads available, then this new task will be started immediately by one of them; otherwise, it's a promise to come back later.

Thus during execution, we first use one thread, then two, then three and so on until all threads are busy. In other words, there will be something of a load balance issue near the beginning of execution, just as we noted earlier for Mergesort.

There are various possible refinements, such as the barrier-like **taskwait** clause.

10.5 Sorting in Rth

Unfortunately, none of the algorithms above is embarrassingly parallel, and most require considerable movement of data. This makes it difficult to code them efficiently in pure R. Fortunately Thrust provides us with a C++ solution, to which the **rthsort()** in **Rth** provides a convenient interface.

And remember, the Bucket Sort above was for multicore platforms. A major bonus of **Rth** is that we can run our code on either multicore machines or GPUs.

Let's test it first:

```
> library(Rth)
Loading required package: Rcpp
> x <- runif(25)
> x
 [1] 0.90189818 0.68357514 0.93200351 0.41806736
 0.40033254 0.09879424
 [7] 0.70001364 0.01025429 0.30682519 0.74398691
 0.04592790 0.57226260
[13] 0.66428642 0.14953737 0.30014257 0.92142903
0.99587218 0.16254603
[19] 0.36737230 0.46898850 0.76138804 0.67405064
0.15926002 0.19043531
[25] 0.81125042
> rthsort(x)
 [1] 0.01025429 0.04592790 0.09879424 0.14953737
 0.15926002 0.16254603
 [7] 0.19043531 0.30014257 0.30682519 0.36737230
 0.40033254 0.41806736
[13] 0.46898850 0.57226260 0.66428642 0.67405064
0.68357514 0.70001364
[19] 0.74398691 0.76138804 0.81125042 0.90189818
0.92142903 0.93200351
[25] 0.99587218
```

Timings will be presented shortly.

Again, one does not need to know Thrust/C/C++/CUDA to use **Rth**, but it is instructive to look at the implementation:

```
// Rth interface to Thrust sort

#include <thrust/device_vector.h>
#include <thrust/sort.h>

#include <Rcpp.h>
#include "backend.h"


RcppExport SEXP rthsort_double(SEXP a,
```

```
  SEXP decreasing, SEXP inplace, SEXP nthreads)
{
  // construct a C++ proxy for the R vector
  Rcpp::NumericVector xa(a);

  // set up device vector and copy xa to it
  thrust::device_vector<double> dx(xa.begin(), xa.end());

  // sort, then copy back to R vector
  if (INTEGER(decreasing)[0])
    thrust::sort(dx.begin(), dx.end(),
       thrust::greater<double>());
  else
    thrust::sort(dx.begin(), dx.end());
  if (INTEGER(inplace)[0]) {
     thrust::copy(dx.begin(), dx.end(), xa.begin());
     // return xa;
  } else {
     Rcpp::NumericVector xb(xa.size());
     thrust::copy(dx.begin(), dx.end(), xb.begin());
     return xb;
  }
}
```

Note that R modes **double** and **integer** are treated separately, so the above needs to be adapted for the latter mode.

An optional argument to **rthsort()**, **decreasing** specifies a sort in descending order if TRUE. Connected with that is use of the **INTEGER()** function. This is from R internals, not **Rcpp**, but it is similar to the latter. We input a SEXP and form a C++ proxy for it. In this case, the input is an integer vector (of length 1).[1] The output of **INTEGER()** is a C++ integer vector, and we want the first (and only) element, which has index 0 in the C++ world.

This is then used in the statement

```
thrust::sort(dx.begin(),dx.end(),thrust::greater<int>());
```

Thrust has an optional argument that specifies the type of comparison to be used in a sort. Here we specify the built-in Thrust function

```
thrust::greater<int>(x,y)
```

[1] Logical values in R can be taken to be integers.

which returns **true** if x > y. Due to the way Thrust's sort function is set up, this amounts to specifying a descending-order sort. But we could set up a custom sort, with our own comparison function tailored to our application.

The fact that **Rcpp** and Thrust are both modeled after C++ STL makes them rather similar to each other; learn one, and then learning the other is much easier. Thus statements such as

```
thrust::copy(dx.begin(), dx.end(), xb.begin());
```

are straightforward.

10.6 Some Timing Comparisons

Here we compare R's built-in serial **sort()**, our **multicore** implementation of Bucket Sort, and **Rth**'s **rthsort()** with an OpenMP backend.[2]

n	function	# threads	time (s)
50000000	**sort()**	1	11.132
50000000	**mcbsort()**	4	11.173
50000000	**rthsort()**	4	4.162
50000000	**mcbsort()**	8	11.009
50000000	**rthsort()**	8	3.852
100000000	**sort()**	1	21.445
100000000	**mcbsort()**	4	21.856
100000000	**rthsort()**	4	7.647
100000000	**mcbsort()**	8	22.137
100000000	**rthsort()**	8	7.728
250000000	**sort()**	1	90.079
250000000	**mcbsort()**	4	51.951
250000000	**rthsort()**	4	22.865
250000000	**mcbsort()**	8	51.005
250000000	**rthsort()**	8	19.806

Moving from R to C/C++ really paid off here. Note too that the pure R parallelization gave no speed advantage until we tried the largest size, $n = 250000000$. And even then, **mcbsort()** did not benefit when we increased the number of threads from 4 to 8.

[2]The directly OpenMP Quicksort in Section 10.4 was not included, as it was not called from R.

A run with $n = 50000000$ on a GeForce GTX 550 Ti GPU took only 1.243 seconds, less than a third of the 8-core run on the multicore machine. However, an attempt with $n = 100000000$ failed, due to the graphics card having insufficient memory. For solutions, see Section 10.7.

10.7 Sorting on Distributed Data

Recall Chapter 9, in which it was explained that with really large data sets, it may be desirable to store a file in chunks. Though it appears as a single file to the user, the file is broken down into many separate files, under different (typically numbered) file names, possibly on different disks and maybe even in geographically separate locations. The Hadoop Distributed File System exemplifies this approach.

In such a situation, our view of sorting is typically different from what we have seen earlier in this chapter. Our input data may be in physically separate files, or possibly in memory but in separate machines, say in a cluster — and we may wish our output to have the same distributed structure. In other words, we would have our sorted array stored in chunked form, with different chunks in different files or on different machines. In this section, we take a look at how this might be done.

The issues of network latency and bandwidth, discussed in Section 2.5, become especially acute in clusters and in distributed data sets. Since sorting is not a parallel operation, these problems are central to developing efficient parallel sorting in such contexts. This is very dependent on one's specific communications context, and only some general suggestions will be made here.

10.7.1 Hyperquicksort

For simplicity, say we have 2^k processes, with ID numbers $0, 1, 2, ..., 2^k - 1$, and that our data is distributed among the processes. We won't present the details here, but roughly it works as follows.

There are k iterations. At the ith one, the processes are broken down into disjoint groups called i-cubes, each consisting of 2^i processes, and with each process being assigned a partner within its i-cube. One process in the i-cube broadcasts its median to all other processes in the i-cube, essentially serving as a pivot. Then each process does a compare-and-exchange operation with its partner, with numbers smaller than the pivot being transferred to the lower-ID partner and the larger numbers going to the higher-ID partner.

When all the dust clears, the vector is in sorted form, though again in a distributed manner across the processes.

10.8 Further Reading

There is a vast literature on parallel sorting and parallel versions of related operations. A very accessible treatment is *Algorithms Sequential and Parallel*, by R. Miller and L. Boxer, Prentice-Hall, 2000.

In reading about good sorting algorithms in the research literature, the reader should beware the various "sorting competitions," such as `http://sortbenchmark.org`. While these results are interesting and shed some light on the problem, keep in mind that the competitions are typically conducted under very idealized conditions, including special hardware. Timings in real life will not be nearly so fast.

Chapter 11

Parallel Prefix Scan

Prefix scan computes cumulative operations, like R's **cumsum()** for cumulative sums:

```
> x <- c(12,5,13)
> cumsum(x)
[1] 12 17 30
```

The scan for sums of (12,5,13) would then be

$$(12, 12 + 5, 12 + 5 + 13) = (12, 17, 30),$$

as we saw above.

11.1 General Formulation

In its general, abstract form, we have some associative operator, $\otimes$, and prefix scan inputs sequence of objects $(x_0, ..., x_{n-1})$, and outputs $(s_0, ..., s_{n-1})$, where

$$\begin{aligned} s_0 &= x_0, \\ s_1 &= x_0 \otimes x_1, \\ s_2 &= x_0 \otimes x_1 \otimes x_2, \\ &..., \\ s_{n-1} &= x_0 \otimes x_1 \otimes ... \otimes x_{n-1} \end{aligned} \tag{11.1}$$

The operands x_i need not be numbers. For instance, they could be matrices, with $\otimes$ being matrix multiplication.

The form of scan used above is called **inclusive** scan, in which x_i is included in s_i. The **exclusive** version omits x_i. So for instance the exclusive cumulative sum vector in the little example above would be

$$(0, 12, 12 + 5) = (0, 12, 17),$$

11.2 Applications

Prefix scan has become a popular tool with which to implement parallel computing algorithms, applicable in a surprising variety of situations. Consider for instance a parallel *filter* operation, like

```
> x
 [1] 19 24 22 47 27  8 28 39 23  4 43 11 49 45 43
 2 13  8 50 41 24 13  7 14 38
> y <- x[x > 28]
> y
[1] 47 39 43 49 45 43 50 41 38
```

With an eye toward parallelizing this operation, let's see how to cast it as a prefix scan problem, as follows:

```
> b <- as.integer(x > 28)
> b
 [1] 0 0 0 1 0 0 0 1 0 0 1 0 1 1 1 0 0 0 1 1 0 0 0 0 1
> cumsum(b)
 [1] 0 0 0 1 1 1 1 2 2 2 3 3 4 5 6 6 6 6 7 8 8 8 8 8 9
```

Look where the vector **b** changes values — at elements 4, 8, 11, 13, 14, 15, 19, 20 and 25. But these are precisely the elements of **x** that go into **y**.

So here **cumsum()**, a prefix scan operation, enabled the filtering operation. Thus, if we can find a way to parallelize prefix scan, we can parallelize filtering. (The creation of **b** above, and the operation of checking changed values in **b**, are embarrassingly parallel.)

And, surprisingly, that gives us an efficient way to parallelize quicksort. The partitioning step — finding all elements less than the pivot and all greater than it — is just two filters, after all. The first step in a bucket sort is also a filter.

11.3 General Strategies

So, how can we parallelize prefix scan? Actually, there are pretty good methods for this.

11.3.1 A Log-Based Method

A common method for parallelizing prefix scan first works with adjacent pairs of the x_i, then pairs spaced two indices apart, then four, then eight, and so on.

For the time being, we'll assume we have n threads, i.e., one for each datum. Clearly this condition will typically not hold, but we'll extend things later. Thread i will handle assignments to x_i. Here's the basic idea, say for $n = 8$:

Step 1:

$$x_1 \leftarrow x_0 + x_1 \tag{11.2}$$
$$x_2 \leftarrow x_1 + x_2 \tag{11.3}$$
$$x_3 \leftarrow x_2 + x_3 \tag{11.4}$$
$$x_4 \leftarrow x_3 + x_4 \tag{11.5}$$
$$x_5 \leftarrow x_4 + x_5 \tag{11.6}$$
$$x_6 \leftarrow x_5 + x_6 \tag{11.7}$$
$$x_7 \leftarrow x_6 + x_7 \tag{11.8}$$

Step 2:

$$x_2 \leftarrow x_0 + x_2 \tag{11.9}$$
$$x_3 \leftarrow x_1 + x_3 \tag{11.10}$$
$$x_4 \leftarrow x_2 + x_4 \tag{11.11}$$
$$x_5 \leftarrow x_3 + x_5 \tag{11.12}$$
$$x_6 \leftarrow x_4 + x_6 \tag{11.13}$$
$$x_7 \leftarrow x_5 + x_7 \tag{11.14}$$

Step 3:

$$
\begin{aligned}
x_4 &\leftarrow x_0 + x_4 & (11.15)\\
x_5 &\leftarrow x_1 + x_5 & (11.16)\\
x_6 &\leftarrow x_2 + x_6 & (11.17)\\
x_7 &\leftarrow x_3 + x_7 & (11.18)
\end{aligned}
$$

In Step 1, we look at elements that are 1 apart, then Step 2 considers the ones that are 2 apart, then 4 for Step 3.

Why does this work? Consider how the contents of x_7 evolve over time. Let a_i be the original x_i, i = 0,1,...,n-1. Then here is x_7 after the various steps:

Step	contents
1	$a_6 + a_7$
2	$a_4 + a_5 + a_6 + a_7$
3	$a_0 + a_1 + a_2 + a_3 + a_4 + a_5 + a_6 + a_7$

So in the end x_7 will indeed contain what it should. Working through the case $i = 2$ shows that x_2 eventually contains $a_0 + a_1 + a_2$, again as it should. Moreover, "eventually" comes early in this case, at the end of Step 2; this will be an important issue below.

For general n, the routing is as follows. At Step i, each x_j is routed both to itself and to $x_{j+2^{i-1}}$, for $j >= 2^{i-1}$.

There will be $log_2 n$ steps if n is a power of 2; otherwise the number of steps is $\lfloor log_2 n \rfloor$.

Note these important points:

- The location x_i appears both as an input and an output in the assignment operations above. In our code implementation, we need to take care that the location is not written to before its value is read.

  One way to do this is to set up an auxiliary array y_i. In odd-numbered steps, the y_i are written with the x_i as inputs, and vice versa for the even-numbered steps. This class of approaches in general—in which we maintain two data objects instead of one, alternating between them—are called *red and black methods*, inspired by a checkerboard, in which adjoining squares are of different colors. Here the x_i are "red" and the y_i are "black."

- Again note the fact that as times goes on, more and more threads become idle; x_i will not change after Step i at the latest, typically earlier. Thus thread i will become idle, and load balancing is poor.
- Synchronization at each step incurs overhead in a multicore/multiprocessor setting. (Worse for GPU if multiple blocks are used).

Now, what if n is greater than p, our number of threads, the typical case? One approach would be to make the assignment of threads to data be dynamic, reconfigured at each step. If at any given step we have k nonidle x_i, then we assign each thread to handle about k/p of the x_i positions.

11.3.2 Another Way

Instead of the above bottom-up approach, we could go top-down, with only one level, as follows. As you've seen before, the natural approach would be to break the vector into chunks, run the algorithm on each chunk, and somehow combine the results:

(a) Break the x_i into p chunks, of size approximately n/p.

(b) Have each thread compute the prefix scan for its chunk.

(c) Compute the prefix scan of the right-hand endpoints of the chunks. (Actually, we need only the first $p-1$.)

(d) Have each thread adjust its own prefix scan according to the result of step (c).

Here's pseudocode for an approach along these lines. Let Ti denote thread i.

```
break the array into p blocks
parallel for i = 0,...,p-1
   Ti does scan of block i, resulting in Si
form new array G of rightmost elements of each Si
do parallel scan of G
parallel for i = 1,...,p-1
   Ti adds Gi to each element of block i+1
```

For example, say we have the array

2 25 26 8 50 3 1 11 7 9 29 10

and wish to compute a sum scan, i.e., cumulative sums. Suppose we have three threads. We break the data into three sections,

2 25 26 8	50 3 1 11	7 9 29 10

and then apply a scan to each section:

2 27 53 61	50 53 54 65	7 16 45 55

But we still don't have the scan of the array overall. That 50, for instance, should be 61+50 = 111 and the 53 should be 61+53 = 114. In other words, 61 must be added to that second section, (50,53,54,65), and 61+65 = 126 must be added to the third section, (7,16,45,55). This then is the last step, yielding

2 27 53 61	111 114 115 126	133 142 171 181

11.4 Implementations of Parallel Prefix Scan

The above pseudocode is easy to implement, and indeed we'll do so in an example below. But it's worth noting that parallel prefix scan is already implemented in various libraries, including:

- The MPI standard actually includes built-in parallel prefix functions, **MPI_Scan()**. A number of choices are offered for the $\otimes$ operator, such as maximum, minimum, sum and product.
- The Thrust library for CUDA or OpenMP/TBB includes functions **inclusive_scan()** and **exclusive_scan()**. We'll see an example in Section 11.6.
- TBB itself offers the **parallel_scan()** function.
- The CUDPP (CUDA Data Parallel Primitives Library) package contains CUDA functions for sorting and other operations, many of which are based on parallel scan.

Some of these are rather complicated, but offer wide generality.

Note carefully that prefix scan is generally an $O(n)$ operation. Thus the discussion in Section 2.9 suggests that communications overhead can be a big issue. The availability of libraries does not necessarily imply that they will work well on your application, on your platform.

11.5 Parallel cumsum() with OpenMP

Here we will write C++ code to do parallel computation of cumulative sums, using OpenMP. The code will be callable from R, via **Rcpp**.

```
// parallel analog of cumsum(), using OMP

#include <Rcpp.h>
#include <omp.h>

// input vector x, number of desired threads nth
RcppExport SEXP ompcumsum(SEXP x, SEXP nth)
{
  Rcpp::NumericVector xa(x);
  int nx = xa.size();

  // cumulative sums vector
  double csms[nx];

  // set number of threads, and allocate adjustment
  // values vector; INTEGER is a SEXP construct
  int nthreads = INTEGER(nth)[0];
  omp_set_num_threads(nthreads);
  // space to store the block endpoint sums
  double adj[nthreads-1];

  int chunksize = nx / nthreads;

  // output vector
  Rcpp::NumericVector csout(nx);

  #pragma omp parallel
  {
     int me = omp_get_thread_num();
     int mystart = me * chunksize,
         myend = mystart + chunksize - 1;
     if (me == nthreads-1) myend = nx - 1;
     int i;

     // do cumulative sums for my chunk
     double sum = 0;
     for (i = mystart; i <= myend; i++) {
        sum += xa[i];
```

```
         csms[i] = sum;
      }

      // find adjustment values
      //
      // first, make sure all the chunk cumsusm
      // are ready
      #pragma omp barrier
      // only one thread need compute and
      // accumulate the right-hand endpoints
      #pragma omp single
      {
         adj[0] = csms[chunksize-1];
         if (nthreads > 2)
            for (i = 1; i < nthreads-1; i++) {
               adj[i] =
                  adj[i-1] + csms[(i+1)*chunksize-1];
            }
      }
      // implied barrier at the end of any
      // 'single' pragma

      // do my adjustments
      double myadj;
      if (me == 0) myadj = 0;
      else myadj = adj[me-1];
      for (i = mystart; i <= myend; i++)
         csout[i] = csms[i] + myadj;
   }
   // implied barrier at the end of any
   // 'parallel' pragma
   return csout;
}
```

The code is a straightforward application of the OpenMP principles we learned earlier. Compile and run as in Section 5.5.5. Note that one must use **as.integer()** for **nth** in the R call, e.g.

```
.Call("ompcumsum",x,as.integer(2))
```

11.5.1 Stack Size Limitations

In the above code, consider the innocuous-looking lines,

```
double csms[nx];
...
Rcpp::NumericVector csout(nx);
```

Recall the discussion in Section 4.1.2, regarding how local variables are usually stored in memory: Each thread is assigned space in memory called a *stack*, in which local variables for a thread are stored. In our example above, **csms** and **csout** are such variables.

The significance of this is that the operating system typically places a limit on the size of a stack. Since our cumulative sum code is typically going to run on very large vectors (otherwise the serial version, R's **cumsum()**, is fast enough), we run the risk of running out of stack space, causing an execution error.

Typical OSs allow you to change the default stack size, in various ways. This will be done in the next section. However, it brings up an issue as to whether we want to follow basic R philosophy of not having *side effects.* In our setting here, if we were willing to violate that informal rule, we could write the above code so that **csout** is one of the arguments to **ompcumsum()**, rather than being its return value. As long as our actual **csout** is a top-level variable, i.e., set at the > command-line level, it would not be on a stack, hence would not cause stack issues.

11.5.2 Let's Try It Out

I ran this code from a C shell, and first set a large stack space of over 4 billion bytes, to accommodate calculation of cumulative sums on an array of length 500 million::

```
% limit stacksize 4000m
```

In the bash shell I would have used **ulimit**, say

```
% ulimit -s 4000000
```

Or, on any system, I could have set the stack size as one of the arguments to **gcc**.

I tried 2 through 16 cores, spaced 2 apart, on the 16-core machine described in this book's Preface. In order to reduce sampling variation, I performed

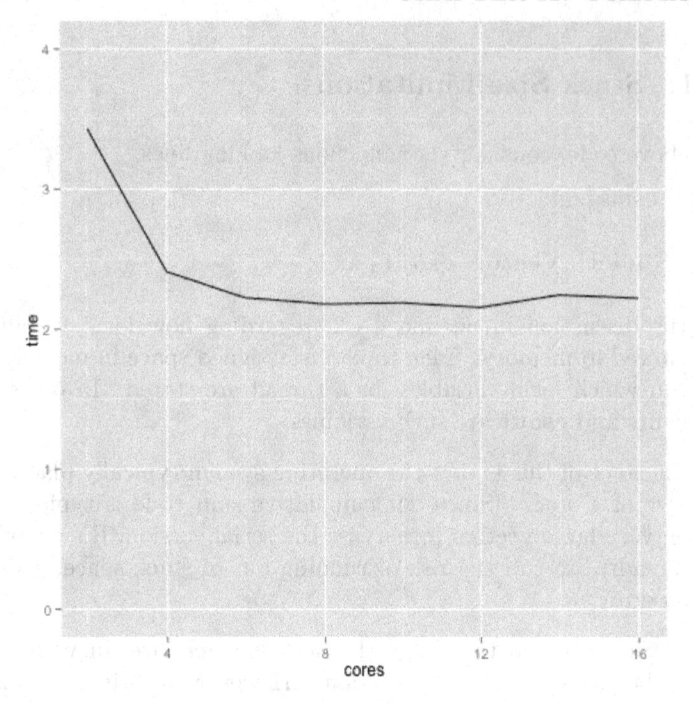

Figure 11.1: OMP Cumulative Sum Run Times

three runs at each value of the number of cores. The results are showing in Figure 11.1.

For this particular sample size, adding more cores increases speed up about 8 cores, after which there is no improvement.

By comparison, the median time for R's **cumsum()** in three runs was 10.553. So, even 2 cores yielded a speedup of much more than 2, reflecting the fact that we are working purely at the C++ level.

11.6 Example: Moving Average

A *moving average* is defined as follows. With input $x_1, .., x_n$ and window width w, the output is $a_w, ..., a_n$, where

$$a_i = \frac{x_{i-w+1} + ... + x_i}{w} \tag{11.19}$$

The goal is to address the question, "What has the recent trend been?" at time i, $i = w, ..., n$.

11.6.1 Rth Code

Rth includes the function **rthma()** to compute moving average in parallel, using an algorithm adapted from the Thrust examples package. Here is the C++ code, contents of the file **rthma.cpp**:

```
// Rth interface to Thrust moving-average example,
// simple_moving_average.cu in
// github.com/thrust/thrust/blob/master/examples/

// C++ code adapted from example in Thrust docs

#include <thrust/device_vector.h>
#include <thrust/scan.h>
#include <thrust/transform.h>
#include <thrust/functional.h>
#include <thrust/sequence.h>
#include <Rcpp.h>
#include "backend.h"  // from Rth

// update function, to calculate current moving
// average value from the previous one
struct minus_and_divide :
   public thrust::binary_function<double,double,
      double>
{
    double w;
    minus_and_divide(double w) : w(w) {}
    __host__ __device__
    double operator()(const double& a,
       const double& b) const
    { return (a - b) / w; }
};

// computes moving averages from x of window width w
RcppExport SEXP rthma(SEXP x, SEXP w, SEXP nthreads)
{
   Rcpp::NumericVector xa(x);
   int wa = INTEGER(w)[0];
```

```
   #if RTH_OMP
   omp_set_num_threads(INT(nthreads));
   #elif RTH_TBB
   tbb::task_scheduler_init init(INT(nthreads));
   #endif

   // set up device vector and copy xa to it
   thrust::device_vector<double> dx(xa.begin(),
      xa.end());

   int xas = xa.size();
   if (xas < wa)
       return 0;

   // allocate device storage for cumulative sums,
   // and compute them
   thrust::device_vector<double>
      csums(xa.size() + 1);
   thrust::exclusive_scan(dx.begin(), dx.end(),
      csums.begin());
   // need one more sum at (actually past) the end
   csums[xas] = xa[xas-1] + csums[xas-1];

   // compute moving averages from cumulative sums
   Rcpp::NumericVector xb(xas - wa + 1);
   thrust::transform(csums.begin() + wa, csums.end(),
      csums.begin(), xb.begin(),
         minus_and_divide(double(wa)));

   return xb;
}
```

11.6.2 Algorithm

Again with the x_i as inputs, it first computes the cumulative sums c_i, using the Thrust function **exclusive_scan()**:

```
thrust::exclusive_scan(dx.begin(), dx.end(),
   csums.begin());
```

Here **dx** contains a copy of the x_i on the device (GPU or OpenMP/TBB), and **csums** will contain our cumulative sums c_i.

Since the numerator of (11.19) is

$$x_{i-w+1} + ... + x_i = c_i - c_{i-w} \tag{11.20}$$

We then need only compute these differences $c_i - c_{i-w}$and divide by w.

To do all this, we use Thrust's **transform()** function:

```
thrust::transform(csums.begin() + wa, csums.end(),
   csums.begin(), xb.begin(),
   minus_and_divide(double(wa)));
```

As the name implies, **transform()** takes one or more inputs, applies a user-specified transformation, and writes to an output vector. You can see that the first two arguments are first the c_i, shifted left by w, and then the c_i themselves. The functor computes the values in (11.20) and divides by w:

```
struct minus_and_divide :
   public thrust::binary_function<double,
      double,double>
{
    double w;
    minus_and_divide(double w) : w(w) {}
    __host__ __device__
    double operator()(const double& a, const double& b)
       const
    { return (a - b) / w; }
};
```

11.6.3 Performance

I ran the code on the 16-core multicore machine again. Since this machine has hyperthread degree 2 (Section 1.4.5.2), it may continue to give speedups through 32 threads. For each number of threads, the experiment consisted of first running the function **runmean()** ("running mean," i.e., moving average) from the R package **caTools** in order to establish a baseline run time:

```
> n <- 1500000000
```

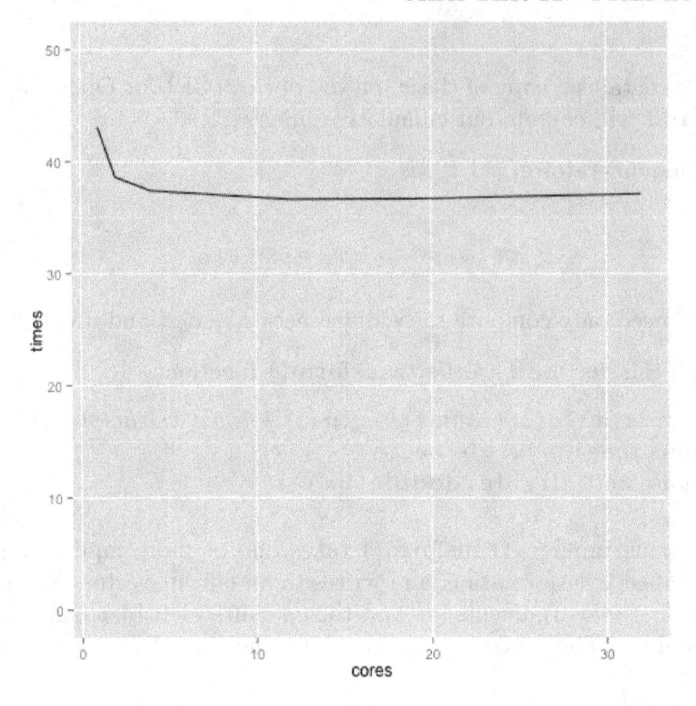

Figure 11.2: Rthma Run Times

```
> x <- runif(n)
> system.time(runmean(x,10))
   user  system elapsed
 39.326  15.925  55.334
```

I then ran **rthma()** for various numbers of cores. Note that I had to choose between OpenMP and TBB, and first selected the former. Look at the results in Figure 11.2.

Again owing to using C++, we already get a performance gain even with a single core, compared to using **runmean()**.

As the number of cores increases, we attain a modest improvement, up through about 12 cores. After that it is flat, possibly with some deterioration as seen for instance in Figure 2.2. A more thorough study would require multiple run times at each value of the number of cores.

Well, what about TBB? Running **rthma()** with a TBB backend, and allowing TBB to choose the number of cores for us, I had a run time of 34.233 seconds, slightly better than the best OpenMP time.

In addition, I tried it on a GPU. Due to memory limitations on the GPU, the test case was smaller. Here are the results:

```
> n <- 250000000; x <- runif(n);
> system.time(runmean(x,10));
   user  system elapsed
  6.972   3.440  10.449
> system.time(.Call("rthma",x,as.integer(10),
    as.integer(-1)))
   user  system elapsed
  2.628   0.724   3.583
```

This is about triple speedup, very nice, though not commensurate with the large number of cores in the GPU. Presumably the communications overhead here was very much a limiting factor.

11.6.4 Use of Lambda Functions

If you have a compiler that allows C++11 *lambda functions*, these can make life much simpler for you if you use Thrust, TBB or anything that uses functors. Let's see how this would work with our C++ function **rthma()** above (scroll down to "changed code").

```
#include <thrust/device_vector.h>
#include <thrust/scan.h>
#include <thrust/transform.h>
#include <thrust/functional.h>
#include <thrust/sequence.h>
#include <Rcpp.h>
#include "backend.h"  // from Rth

// struct minus_and_divide now deleted

RcppExport SEXP rthma(SEXP x, SEXP w, SEXP nthreads)
{
   Rcpp::NumericVector xa(x);
   int wa = INTEGER(w)[0];
   #if RTH_OMP
   omp_set_num_threads(INT(nthreads));
   #elif RTH_TBB
   tbb::task_scheduler_init init(INT(nthreads));
   #endif
   thrust::device_vector<double> dx(xa.begin(),
```

```
      xa.end());
   int xas = xa.size();
   if (xas < wa) return 0;
   thrust::device_vector<double>
      csums(xa.size() + 1);
   thrust::exclusive_scan(dx.begin(), dx.end(),
      csums.begin());
   csums[xas] = xa[xas-1] + csums[xas-1];
   Rcpp::NumericVector xb(xas - wa + 1);

   // changed code
   thrust::transform(csums.begin() + wa, csums.end(),
      csums.begin(), xb.begin(),
      // lambda function
      [=] (double& a, double& b) { return( (a-b)/wa);
});

   return xb;
}
```

Just what is going on in the line

```
[=] (double& a, double& b) { return( (a-b)/wa); });
```

We are creating a function object here, as we did in the earlier version with an **operator()** notation within a **struct**. But here we are doing so right on the spot, in one of the arguments to **thrust::transform()**. It's similar to the concept of *anonymous* functions in R. Here are the details:

- The brackets in [=] tells the compiler that a function object is about to begin. (The = sign will be explained shortly.)

- The variable **wa** that was local to the function **rthma()** is considered "global" to our lambda function, thus accessible to it *without passing* **wa** *as an argument to that function.* (This too is similar to the R case.) We say that **wa** is *captured* by that function.

  The = indicates that we wish to use **wa** *by value*, i.e., just for its value alone, rather than it being a pointer that we wish to reference. If it had been the latter (with the capability of changing the pointed-to location), we'd use & instead of =.

- Here **a** and **b** are ordinary arguments.

All this is so much clearer and cleaner than using a functor!

By the way, in using **g++** for the compilation, I did need to add the **-std=c++11** compiler flag to use lambda functions.

Chapter 12

Parallel Matrix Operations

The matrix is one of the core types in the R language. And in recent years, the range of applications of matrices has expanded, from the traditional areas such as regression analysis and principal components to image processing and the analysis of random graphs in social networks.

In modern applications, matrices are often huge, with sizes of tens of thousands of rows, and far more, being commonplace. Thus there is a large demand for parallel matrix algorithms, the subject of this chapter.

There is a plethora of parallel matrix software available to R programmers. For GPU, for instance, there are R packages **gputools**, **gmatrix** (this one especially noteworthy because it enables us to follow the Principle of "Leave It There," Section 2.5.2), **MAGMA** and so on. Indeed, more and more of them are being developed. See also PLASMA and **HiPLAR**, as well as **pdbDMAT**.

Thus this chapter cannot provide comprehensive coverage of everything that is out there for parallel linear algebra in R. In addition, in some special cases the libraries don't quite do what we need. Thus, our coverage will consist mainly of basic, general principles, with examples using some of the libraries. Selection of libraries for the examples will be based mainly on ease of installation and use.

12.1 Tiled Matrices

Parallel processing of course relies on finding a way to partition the work to be done. In the matrix algorithms case, this is typically done by dividing a matrix into blocks, often called *tiles* these days.

For example, let

$$A = \begin{pmatrix} 1 & 5 & 12 \\ 0 & 3 & 6 \\ 4 & 8 & 2 \end{pmatrix} \tag{12.1}$$

and

$$B = \begin{pmatrix} 0 & 2 & 5 \\ 0 & 9 & 10 \\ 1 & 1 & 2 \end{pmatrix}, \tag{12.2}$$

so that

$$C = AB = \begin{pmatrix} 12 & 59 & 79 \\ 6 & 33 & 42 \\ 2 & 82 & 104 \end{pmatrix}. \tag{12.3}$$

We could partition A as

$$A = \begin{pmatrix} A_{11} & A_{12} \\ A_{21} & A_{22} \end{pmatrix}, \tag{12.4}$$

where

$$A_{11} = \begin{pmatrix} 1 & 5 \\ 0 & 3 \end{pmatrix}, \tag{12.5}$$

$$A_{12} = \begin{pmatrix} 12 \\ 6 \end{pmatrix}, \tag{12.6}$$

$$A_{21} = \begin{pmatrix} 4 & 8 \end{pmatrix} \tag{12.7}$$

and

$$A_{22} = \left(\begin{array}{c} 2 \end{array} \right). \tag{12.8}$$

Similarly we would partition B and C into blocks of a compatible size to A ("compatible" to be explained shortly),

$$B = \left(\begin{array}{cc} B_{11} & B_{12} \\ B_{21} & B_{22} \end{array} \right) \tag{12.9}$$

and

$$C = \left(\begin{array}{cc} C_{11} & C_{12} \\ C_{21} & C_{22} \end{array} \right), \tag{12.10}$$

with for example

$$B_{21} = \left(\begin{array}{cc} 1 & 1 \end{array} \right). \tag{12.11}$$

The key point is that multiplication still works if we pretend that those submatrices are numbers! The matrices A, B and C would all be thought of as of "2×2" size, in which case we would have for example

$$C_{11} = A_{11}B_{11} + A_{12}B_{21}, \tag{12.12}$$

(In that expression on the right, we are still treating A_{11} etc. as matrices, even though we got the equation by pretending they were numbers.)

The reader should verify this really is correct as matrices, i.e., the computation on the right side really does yield a matrix equal to C_{11}. You can see now what kind of compatibility is needed. For instance, the number of rows in B_{11} must match the number of columns in A_{11} and so on.

12.2 Example: Snowdoop Approach

Suppose we wish to compute the matrix product AB. In the Snowdoop context, the multiplier A may be stored in chunks at various nodes. Or, another situation might be one in which we wish to use a GPU but A is too large to fit in the GPU's memory. In that case again we may wish to process A in chunked form. Note that this is a special case of tiling. Here we will see how to deal with this, using the GPU case for illustration.

Consider computation of a product Ax, where A is a matrix and x is a conformable vector. Suppose A is too large to fit in our GPU. Our strategy will be simple: Break A into tiles of rows, multiply x by each tile of A, and concatenate the results to yield Ax, as we did in Section 1.4.4. The code is equally simple, say with **gputools** (Section 6.7.1):

```
# GPUTiling.R

biggpuax <- function(a,x,ntiles)
{
   require(parallel)
   require(gputools)
   nrx <- nrow(a)
   y <- vector(length=nrx)
   tilesize <- floor(nrx / ntiles)
   for (i in 1:ntiles) {
      tilebegin <- (i-1) * tilesize + 1
      tileend <- i * tilesize
      if (i == ntiles) tileend <- nrx
      tile <- tilebegin:tileend
      y[tile] <- gpuMatMult(a[tile,,drop=FALSE],x)
   }
   y
}
```

Note that in this (single) GPU context, although the work within a tile is done in parallel, it is still serial between tiles. Moreover, we incur the overhead of communication between the CPU and GPU. This may hamper our efforts to achieve speedup.

On the other hand, if we have multiple GPUs, or if we have a Snowdoop setting, this partitioning could be a win.

12.3 Parallel Matrix Multiplication

Since so many parallel matrix algorithms rely on matrix multiplication, a core issue is how to parallelize that operation, generalizing the tiled approach in the last section.

As multiplication is "embarrassingly parallel," one might at first think it is easy to do efficiently. However, there are serious overhead issues on any platform, as we'll see.

Denote our desired product by AB. Let's suppose for the sake of simplicity that each of the matrices to be multiplied is of dimensions $n \times n$. Let p denote the number of "processes," such as shared-memory threads or message-passing nodes.

12.3.1 Multiplication on Message-Passing Systems

Sections 1.4.4 and 12.2 showed how to parallelize a matrix-vector product computation in **snow**, by breaking the matrix rows into tiles, and then exploiting the tiling properties of matrices. Computation of matrix-matrix products can be done in the same way. But there are more sophisticated methods available.

12.3.1.1 Distributed Storage

Recall the concept of breaking files into chunks, such as with the Hadoop Distributed File System. Typical algorithms for the message-passing setting assume that the matrices A and B in the product AB are stored in a distributed manner across the nodes, in a cluster, and that the product will be distributed too.

In addition to formal MapReduce settings, this situation could arise for several reasons:

- The application is such that it is natural for each node to possess only part of A and B.
- One node, say node 0, originally contains all of A and B, but in order to conserve network communication time, it sends each node only parts of those matrices.
- An entire matrix would not fit in the available memory at the individual nodes.

12.3.1.2 Fox's Algorithm

For ease of exposition, let's assume that $\sqrt{p}$ evenly divides n, and partition each matrix into tiles of size $\sqrt{p} \times \sqrt{p}$. Each matrix will be divided into m rows and m columns of blocks, where $m = n/\sqrt{p}$.

Consider the node that has the responsibility for calculating block (i,j) of the product C, which it calculates as

$$C_{ij} = A_{i1}B_{1j} + A_{i2}B_{2j} + ... + A_{ii}B_{ij} + ... + A_{i,m}B_{m,j} \tag{12.13}$$

It will be convenient here (only in this section) to number rows and columns starting at 0 instead of 1, as the code uses the mod operator based on that setting. Now rearrange (12.13) with A_{ii} first:

$$A_{ii}B_{ij}+A_{i,i+1}B_{i+1,j}+...+A_{i,m-1}B_{m-1,j}+A_{i0}B_{0j}+A_{i1}B_{1j}+...+A_{i,i-1}B_{i-1,j} \tag{12.14}$$

The algorithm is then as follows. The node which is handling the computation of C_{ij} does this (in parallel with the other nodes which are working with their own values of i and j):

```
iup   = i+1 mod m;
idown = i-1 mod m;
for (k = 0; k < m; k++) {
    km = (i+k) mod m;
    broadcast(A[i,km]) to all nodes handling row i of C;
    C[i,j] = C[i,j] + A[i,km]*B[km,j]
    send B[km,j] to the node handling C[idown,j]
    receive new B[km+1 mod m,j] from the node
       handling C[iup,j]
}
```

The main idea is to have the various computational nodes repeatedly exchange submatrices with each other, timed so that a node receives the submatrix it needs for its computation "just in time."

The algorithm can be adapted in the obvious way to nonsquare matrices, etc.

12.3.1.3 Overhead Issues

In a cluster context, the usual overhead issues of network communication are central.

There is a lot of opportunity here to overlap computation and communication, which is the best way to solve the communication problem. (Recall the concept of *latency hiding*, Section 2.5.)

Obviously this algorithm is best suited to settings in which we have PEs (Section 2.5.1.2) in a *mesh* topology, meaning that each PE is connected to its "north," "south," "east and "west" neighbors, or something similar, even a torus. MPI broadcast operations, used with its advanced communicator capability, can really take advantage of this, for instance.

12.3.2 Multiplication on Multicore Machines

Since a matrix multiplication in serial form consists of nested loops, a natural way to parallelize the operation in OpenMP is through the **for** pragma, e.g.

```
// nrowsa is the number of rows of A etc.
#pragma omp parallel for
for (i = 0; i < nrowsa; i++)
   for (j = 0; j < ncolsb; j++) {
      sum = 0;
      for (k = 0; k< ncolsa; k++)
         sum += a[i][k] * b[k][j];
      c[i][j] = sum;
   }
}
```

This parallelizes the outer loop. Let p denote the number of threads. Since each value of **i** in that loop handles one row of A, what we are doing here is breaking the matrix into p sets of rows of A. A thread computes the product of a row of A and all of B, which becomes the corresponding row of C.

Note, though, that from the discussion on page 122 we know the sets of rows are not necessarily "chunks;" the rows of A processed by a given thread may not be contiguous. We will return to this point shortly.

12.3.2.1 Overhead Issues

First and foremost, cache effects must be considered. (The reader may wish to review Sections 2.3.1 and 5.8 before continuing.) Suppose for the moment that we are working purely in C/C++, which uses row-major storage.

As we execute the innermost ("k") loop above, we are traversing a row of A and a column of B. We thus have fairly good spatial locality for A and C, but poor locality for B. R of course uses column-major storage, but the same considerations apply.

Furthermore, as noted above, the rows executed by any given thread may not be contiguous, again adversely impacting spatial locality. We may wish to remedy this, for example, by using the chunking parameters presented in Section 5.3.

We can't get good locality everywhere, so we might leave it as above, but there are other considerations. One issue is task granularity. As discussed before, e.g. in Section 5.3, with too large a task size we risk having poor load balance near the end of our computation. If we do our tiling only in A, it may be difficult to set a small enough task size.

So, efficient shared-memory software typically also parallelizes work on the columns of A and C above, so that we parallelize not only the "i" loop above but also the "j" one. In OpenMP, one could use the **collapse** clause. For example,

```
#pragma omp for collapse(2)
```

means, "parallelize the 2 nested loops that follow."

There is yet another issue—having *aligned* data objects. Well, you wouldn't want unaligned data, would you? But seriously, it actually is a big issue, as follows.

Let r denote the memory address at which our matrix A starts in memory. If r is not a multiple of the cache block size, say 64 bytes, then A will start in the middle of some memory block. Suppose part of the matrix B is also in that block. Then we may have the *false sharing* problem discussed in Section 5.8. This could put a significant damper on performance. We may be able to get the compiler to align objects for us, or we might simply place some "filler" zeroes at the beginning of A to arrange for the "real" part of A to begin at an address that is a multiple of 64.

As you can see, good parallel computation sometimes requires some low-level tweaking.

12.3.3 Matrix Multiplication on GPUs

Again think of a matrix product $AB = C$. Given that CUDA tends to work better if we use a large number of threads, a natural choice is for each thread to compute one element of the product C, like this:

```
__global__ void matmul(float *a,float *b,float *c,
   int nrowsa,int ncolsa,int ncolsb)
{  int k,i,j; float sum;
   // this thread will compute c[i][j]; the values of
   // i and j will be determined according to thread
   // and block ID (not shown)
   sum = 0;
   for (k = 0; k < ncolsa; k++)
      // add a[i,k] * b[k,j] to sum
      sum += a[i*ncolsa+k] * b[k*ncolsb+j];
   // assign to c[i,j]
   c[i*ncolsb+j] = sum;
}
```

This should produce a good speedup. But we can do even better, much better, as discussed below.

12.3.3.1 Overhead Issues

As is typical on GPUs, memory issues are central. First, if the matrices to be multiplied are not already on the "device," i.e., in the GPU memory, they must be copied there, causing a delay.

Second, note again that large matrices may not fit in the GPU memory. In this case we must resort to tiling as in Section 12.2 (in addition to whatever tiling is done in the GPU computation itself), which means we will incur the copying overhead multiple times.

Another memory issue is *stride* (Section 3.11). A big difference between programming on multicore machines and GPUs is that in the latter case, there is typically available highly detailed information on the hardware structure, such as the GPU memory-interleaving factor. In tweaking our algorithm for maximum performance, we want to design it with a stride that keeps all the banks busy.

Finally, there is the matter of GPU *shared memory*. Recall that the name is misleading; actually this memory is a programmer-managed cache.[1] One must write the code to copy data to and from the GPU global memory to the shared memory, to take advantage of the latter's speed.

[1]The memory is shared among all the threads in a given GPU block.

For instance, here is an excerpt of the CUDA matrix-multiply code presented in a talk given by Prof. Richard Edgar:[2]

```
int a = aBegin, b = bBegin; a <= aEnd; a += aStep,
      b+= bStep) {
   __shared__ float As[BLOCK_SIZE][BLOCK_SIZE];
   __shared__ float Bs[BLOCK_SIZE][BLOCK_SIZE];
   // Load matrices from global memory into shared memory
   // Each thread loads one element of each sub-matrix
   As[ty][tx] = A[a + (dc_wA * ty) + tx];
   Bs[ty][tx] = B[b + (dc_wB * ty) + tx];
```

Here **A** and **B** are in the GPU global memory, and we copy chunks of them into shared-memory arrays **As** and **Bs**. Since the code has been designed so that **As** and **Bs** are accessed frequently during a certain period of the execution, it is worth incurring the copying delay to exploit the fast shared memory.

Fortunately, the authors of the CUBLAS library have already done all the worrying about such matters for you. They have written very finely hand-tuned code for good performance, making good use of GPU shared memory and so on.

12.4 BLAS Libraries

In any discussion of high-performance matrix operations, the first question that arises is, "Which BLAS are you using?"

12.4.1 Overview

BLAS is an acronym for the Basic Linear Algebra Subprograms, a library of functions that perform very low-level operations such as matrix addition and multiplication. As will be discussed below, there are various BLAS implementations, all of them tailored (to varying degrees) to having good cache behavior and to otherwise have good performance.

R uses "vanilla" BLAS, CBLAS, which for instance comes standard with Ubuntu Linux. However, one may build R from source to include one's own favorite BLAS, or to have the choice of BLAS done dynamically each time one runs R.

[2]The same or similar code is available in full on the NVIDIA CUDA Samples website.

Thus even in ordinary serial computation, e.g. R's **%*%** operator, the speed of matrix operations may vary according to the version of BLAS that your implementation of R was built with.

In our context here of parallel computation, one version of BLAS of special interest is OpenBLAS, a multithreaded version that can bring performance gains on multicore machines, and also includes various efficiencies that greatly improve performance even in the serial case. We'll take a closer look at it in Section 12.5.

As noted, there is also CUBLAS, a version of BLAS for NVIDIA GPUs, highly tailored to that platform. A number of R packages, such as **gputools** and **gmatrix**, make use of this library, and of course you can write your own special-purpose R interfaces to it, say using **Rcpp** to interface it to R.

For message-passing systems (clusters or multicore), there is PBLAS, designed to run on top of MPI. The MAGMA library, with the R interface **magma**, aims to obtain good performance on hybrid platforms, meaning multicore systems that also have GPUs.

BLAS libraries also form the basis of libraries that perform more advanced matrix operations, such as matrix inversion and eigenvalue computation. LAPACK is a widely-used package for this. as is ScaLAPACK for PBLAS.

12.5 Example: Performance of OpenBLAS

OpenBLAS is relatively new, having taken over a discontinued project, GotoBLAS. It is not yet clear what its long-term prospects are, but it seems very promising indeed.

For the timing experiments below, I simply switched from the default BLAS to OpenBLAS. This required setting some file permissions, and since I was running on a machine on which I did not have root access, I installed my own copy of R, in a directory **/home/matloff/MyR311**. (I had to build R with the option **–with-shared-blas**.) I also downloaded and built OpenBLAS, installing it in **/home/matloff/MyOpenBLAS**. I then needed to replace the R standard BLAS library via a symbolic link, as follows.

I entered the directory **/home/matloffMyR311/lib/R/lib** and did the operations

```
$ mv libRblas.so libRblas.so.SAVE
$ ln -s /home/matloff/MyOpenBLAS/lib/libopenblas.so \\
    libRblas.so
```

So, now whenever I run R, it loads OpenBLAS instead of the default BLAS.

OpenBLAS is a threaded application. It doesn't use OpenMP to manage threads, but it does allow one to set the number of threads using the OpenMP environment variable, e.g. in the **bash** shell,

```
$ export OMP_NUM_THREADS=2
```

If the number of threads is not set, OpenBLAS will use all available cores.[3] Note, though, that this may not be optimal, as we will see later.

It is important to note that that is all I had to do. From that point on, if I wanted to compute the matrix product AB in *parallel*, I just used ordinary R syntax:

```
> c <- a %*% b
```

I ran a squaring operation of random matrices of size 5000×5000 for 1, 2, 4, 6, 8, 10, 12, 14 and 16 cores on the 16-core machine described in this book's preface. Let's look at the timing, in Figure 12.1. Though there is quite a bit of sampling variability and we would need to do multiple runs for a smoother graph, the results are clear: We are achieving linear speedup (e.g. doubling the number of cores cuts the run time approximately in half) up to about six cores, after which the returns are diminishing, if positive.

Note that not all multicore systems are alike, in terms of how resources are shared, say within groups of cores. More than one core may share a cache, for instance. Thus it's hard to predict at what point the "diminishing returns" effect will occur for any given application and any given hardware platform.

Performance of numerical algorithms is not just about speed; we must also consider accuracy. OpenBLAS derives its speed not just from making use of multiple cores, but also from various tweaks of the code, yielding a very fine degree of optimization. One can thus envision a development team so obsessed with speed that they might cut some corners regarding numerical accuracy. Thus the latter is a subject of legitimate concern.

I conducted a brief experiment to investigate this. I generated $p \times p$ correlation matrices, with all pairs of variables having correlation ρ, using the code

```
covrho <- function(p,rho) {
   m <- diag(p)
```

[3]If the machine has hyperthreading (Section 1.4.5.2), the number of "cores" will be the product of the number of cores and the degree of hyperthreading.

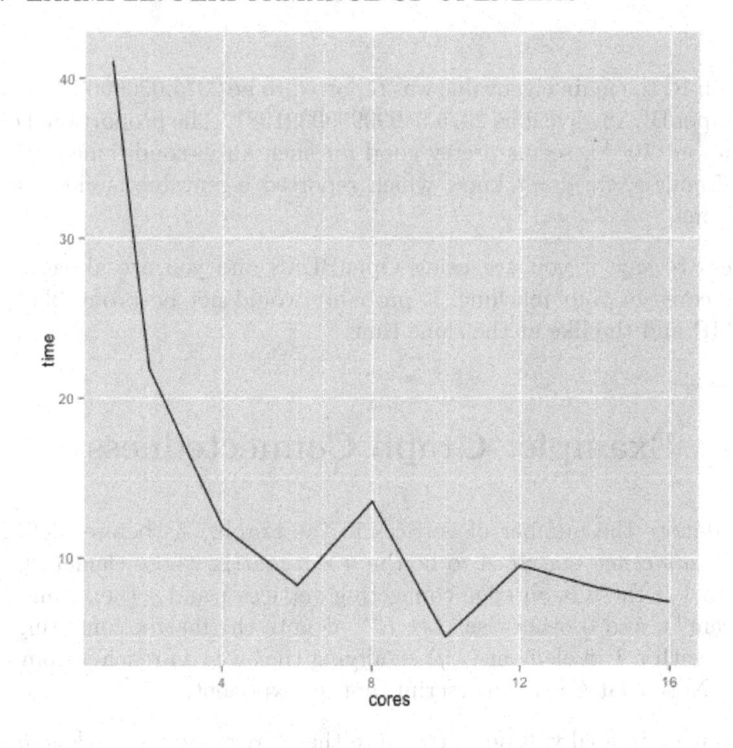

Figure 12.1: OpenBLAS Run Times

```
    m[row(m) != col(m)] <- rho
    m
}
```

The chosen application was eigenvalue computation, using R's **eigen()** function. I ran this on the 16-core machine that has a hyperthreading degree of 2, and did not specify the number of threads, so OpenBLAS likely used 32 threads.

First, one more timing: With $p = 2500$ and $\rho = 0.95$, OpenBLAS handily beat R's stock BLAS, with a time of 12.101 seconds, versus 57.407 for stock BLAS.

Now, for the accuracy test: On the advice from a reader of my blog, Mad (Data) Scientist, I first set an R parameter to display 20 decimal digits in output:

```
options(digits = 20)
```

For plain R, the main eigenvalue was reported to be 2375.050000000147338, while OpenBLAS gave it as 2375.049999999991087. The proportional difference, about 10^{-14}, seems pretty good for such an ill-conditioned matrix.[4] And of course, we don't know which reported eigenvalue is closer to the "true" one.

Needless to say, if you are using OpenBLAS and you are already using all the cores in your machine, it probably would not be profitable to use OpenMP and the like at the same time.

12.6 Example: Graph Connectedness

Let n denote the number of vertices in the graph. As before, define the graph's *adjacency matrix* A to be the $n \times n$ matrix whose element (i,j) is equal to 1 if there is an edge connecting vertices i and j (i.e., i and j are "adjacent"), and 0 otherwise. Let $R^{(k)}$ denote the matrix consisting of 1s and 0s, with a 1 in element (i,j) signifying that we can reach j from i in k steps. (Note that k is a superscript, not an exponent.)

Also, our main goal will be to compute the corresponding *reachability matrix* R, whose (i,j) element is 1 or 0, depending on whether one can reach j in some multistep path from i. In particular, we are interested in determining whether the graph is **connected**, meaning that every vertex eventually leads to every other vertex, which will be true if and only if R consists of all 1s. Let us consider the relationship between the $R^{(k)}$ and R.

12.6.1 Analysis

First, note that

$$R = b(\sum_{k=1}^{n-1} R^{(k)}) \tag{12.15}$$

where **b()** performs a Boolean operation, changing nonzero numbers to 1s and keeping 0s at 0s.

[4] The term means that slight changes in the matrix may result in large changes in output, in this case in the principal eigenvalue. In such a setting, roundoff error could be quite serious.

So, if we calculate all the $R^{(k)}$, then we have R. But there is a trick we can employ to avoid finding them all.

To see this, suppose there exists a path of 16 steps from vertex 8 to vertex 3. If the graph is *directed*, meaning that A is not symmetric, then it may be the case that there is no path from vertex 3 that eventually returns to that node. If so, then

$$R_{83}^{(16)} = 1 \tag{12.16}$$

but

$$R_{83}^{(k)} = 0, \quad k > 16 \tag{12.17}$$

So it would seem that we do indeed need to calculate all the $R^{(k)}$. But we can actually avoid that, by first making a small modifcation to our graph: We add artificial "edges" from every vertex to itself. In matrix terms, this means setting the diagonal elements of A to 1.

Now, in our above example we have

$$R_{83}^{(k)} = 1, \quad k \geq 16 \tag{12.18}$$

since, after step 16, we can keep going from vertex 3 to itself $k - 16$ times.

This saves us a lot of work—we need compute only $R^{(n-1)}$. So, how can we do this? We need one last ingredient:

$$R^{(k)} = b(A^k) \tag{12.19}$$

Why is this true? For instance, think of trying to go from vertex 2 to vertex 7 in two steps. Here are the possibilities:

- go from 2 to 1 and then from 1 to 7 or
- go from 2 to 2 and then from 2 to 7 or
- go from 2 to 3 and then from 3 to 7 or
- ...

But this says that

$$R_{27}^{(2)} = b(a_{21}a_{17} + a_{22}a_{27} + a_{23}a_{37} + ...) \tag{12.20}$$

The key point is that the argument to **b()** here is $(A^2)_{27}$, demonstrating (12.19).

So, the original graph connectivity problem reduces to a matrix power problem. We simply compute A^{n-1} (and then apply **b()**).

12.6.2 The "Log Trick"

Moreover, we can save computation there as well, by using the "log trick." Say we want the 16th power of some matrix B. We can square it, yielding B^2, then square *that*, yielding B^4. Two more squarings then give us B^8 and B^{16}. That would be only four matrix-multiply operations, instead of 15. In general, for power 2^m, we need m steps.

In our graph connectivity setting, we need power $n - 1$, but we might as well go for power

$$2^{\lceil \log_2 n-1 \rceil} \tag{12.21}$$

12.6.3 Parallel Computation

Any parallel vehicle for matrix multiplication could be used for computing the matrix powers, e.g. OpenBLAS or GPUs. In addition, my CRAN package **matpow**, described in the next section, can be used to facilitate the process.

By the way, one of this book's internal reviewers raised the following question: The matrix A likely can be *diagonalized*, i.e., a matrix C can be found such that

$$A = C^{-1}DC \tag{12.22}$$

where D is a diagonal matrix whose elements are the eigenvalues of A (Appendix A). Then

$$A^k = C^{-1}D^kC \tag{12.23}$$

The matrix D^k is trivial to compute. So, if one has a parallel method for finding the eigenvalues of a matrix (which also gives us C), then this would be another parallel method for computing A^k.

However, the finding of eigenvalues is not an embarrassingly parallel operation. And in our graph connectivity application, we need exact numbers, without roundoff.

Actually, one method for computing eigenvalues does involve matrix powers.

12.6.4 The matpow Package

My matrix-powers computation package with Jack Norman, **matpow**, on CRAN, enables flexible, convenient calculation of matrix powers.

12.6.4.1 Features

The **matpow** package is flexible, in two important aspects:

- A user-supplied *callback* function can be specified, which enables application specific actions to be performed at the end of each iteration. For instance, if one is using the power method to find eigenvalues, one can check for convergence after each iteration, and end the iteration process if convergence has been reached.
- Any kind of matrix class/multiplication type can be accommodated—the built-in R "**matrix**" class, GPU multiplication and so on. This means we can easily parallelize the calculation of matrix powers.

12.7 Solving Systems of Linear Equations

Suppose we have a system of equations

$$a_{i1}x_1 + ... + a_{i,n}x_n = b_i, \ i = 1, ..., n, \tag{12.24}$$

where the x_i are the unknowns to be solved for. Such systems occur often in data science, such as in linear regression analysis, computation of Maximum Likelihood Estimates, and so on.

As you know, this system can be represented compactly as

$$Ax = b, \tag{12.25}$$

where A is $n \times n$, with elements a_{ij}, and x and b ares $n \times 1$, with elements b_i.

In principle, we can find x simply by computing $A^{-1}b$. However, we need to consider aspects such as numerical accuracy and the degree to which our methods are "embarrassingly parallel" (or not).

12.7.1 The Classical Approach: Gaussian Elimination and the LU Decomposition

Form the $n \times (n+1)$ matrix $C = (A|b)$ by appending the column vector b to the right of A. Then we work on the rows of C, with the pseudocode for the sequential case in the most basic version being

```
for ii = 1 to n
   divide row ii by c[ii][ii]
   for r = 1 to n, r != ii
      replace row r by row r - c[r][ii] times row ii
```

In the divide operation in the above pseudocode, c_{ii} might be 0, or close to 0. In that case, a *pivoting* operation is performed (not shown in the pseudocode): that row is first swapped with another one further down.

This transforms C to *reduced row echelon form*, in which A is now the identity matrix I and b is now our solution vector x. (If we had done any pivoting, we must rearrange b to the original ordering.)

An important variation is to transform only to *row echelon form*. This means that C ends up in upper triangular form, with all the elements c_{ij} with $i > j$ being 0, and with all diagonal elements being equal to 1. Here is the pseudocode:

```
for ii = 1 to n
   divide row ii by c[ii][ii]
   for r = ii+1 to n  // vacuous if r = n-1
      replace row r by row r - c[r][ii] times row ii
```

This corresponds to a new set of equations,

$$\begin{aligned}
c_{11}x_1 + c_{12}x_2 + c_{13}x_3 + ... + c_{1,n}x_n &= b_1 \\
c_{22}x_2 + c_{23}x_3 + ... + c_{2,n}x_n &= b_2 \\
c_{33}x_3 + ... + c_{3,n}x_n &= b_3 \\
... \\
c_{n,n}x_n &= b_n
\end{aligned}$$

We then find the x_i via *back substitution:*

```
x[n] = b[n] / c[n,n]
for i = n-1 downto 1
   x[i] =
      (b[i]-c[i][n-1]*x[n-1]-...-c[i][i+1]*x[i+1])/c[i][i]
```

Finding the row echelon form is related to the famous LU *decomposition,* which writes A as

$$A = LU$$

where L and U are lower- and upper-triangular matrices, respectively. One can find the inverses of L and U via back substitution, and then get

$$A^{-1} = U^{-1}L^{-1} \tag{12.26}$$

At present, R uses this method for its **solve()** function, which (by the user's choice) either solves $Ax = b$ or simply finds A^{-1}.

All in all, row echelon form should save us some work, as we don't have to repeatedly work on the upper rows of the matrix as in the reduced row echelon form.

With pivoting, the numerical stability with either form is good. But what about parallelizability? This is easy in the reduced row echelon form, as we can statically assign a group of rows to each thread. But in the row echelon form, we have fewer and fewer rows to process as time goes on, making it harder to keep all the threads busy.

12.7.2 The Jacobi Algorithm

One can rewrite (12.24) as

$$x_i = \frac{1}{a_{ii}}[b_i - (a_{i1}x_1 + ... + a_{i,i-1}x_{i-1} + a_{i,i+1}x_{i+1} + ... + a_{i,n}x_n)], \; i = 1, ..., n. \tag{12.27}$$

This suggests a natural iterative algorithm for solving the equations: We start with our guess being, say, $x_i = b_i$ for all i. At our kth iteration, we find our (k+1)st guess by plugging in our kth guess into the right-hand side of (12.27). We keep iterating until the difference between successive guesses is small enough to indicate convergence.

This algorithm is guaranteed to converge if each diagonal element of A is larger in absolute value than the sum of the absolute values of the other elements in its row. But it should work well if our initial guess is near the true value of x.

That last condition may seem unrealistic, but in data science we have many iterative algorithms that require a matrix inversion (or equivalent) at each iteration, such as generalized linear models (**glm()** in R), multiparameter Maximum Likelihood Estimation and so on. Jacobi may work well here for updating the matrix inverse from one iteration to the next.

12.7.2.1 Parallelization

Parallelizing of this algorithm is easy: Just assign each process to handle a section of $x = (x_0, x_1, ..., x_{n-1})$. Note that this means that each process must make sure that all other processes get the new value of its section after every iteration.

To parallelize this algorithm, note that in matrix terms (12.27) can be expressed as

$$x^{(k+1)} = D^{-1}(b - Ox^{(k)}) \tag{12.28}$$

where D is the diagonal matrix consisting of the diagonal elements of A (so its inverse is just the diagonal matrix consisting of the reciprocals of those elements), O is the square matrix obtained by replacing A's diagonal elements by 0s, and $x^{(i)}$ is our guess for x in the ith iteration. This reduces the problem to one of matrix multiplication, and thus we can parallelize the Jacobi algorithm by utilizing a method for doing parallel matrix multiplication.

12.7.3 Example: R/gputools Implementation of Jacobi

Here's the R code, using **gputools**:

```
jcb <- function(a,b,eps) {
   n <- length(b)
   d <- diag(a)  # a vector, not a matrix
   tmp <- diag(d)  # a matrix, not a vector
   o <- a - diag(d)
   di <- 1/d
   x <- b  # initial guess, could be better
   repeat {
      oldx <- x
      tmp <- gpuMatMult(o,x)
      tmp <- b - tmp
      x <- di * tmp  # elementwise multiplication
      if (sum(abs(x-oldx)) < n * eps) return(x)
   }
}
```

12.7.4 QR Decomposition

The famous *QR decomposition* factors our matrix A as

$$A = QR$$

where Q and R are orthogonal and upper-triangular matrices, respectively (Section A.5). This is a favored method for linear regression and eigenvalue problems, due to its numerical stability and serial speed; depending on what quantities are computed, in some problems its time complexity is $O(n^2)$ rather than $O(n^3)$ as in Gaussian elimination, for example.

However, it is difficult to parallelize. Nevertheless, it is the method used in **gputools**' function **gpuSolve()**, the analog of **solve()** from base R.

12.7.5 Some Timing Results

R's **solve()** function, as noted earlier, uses LU decomposition. Timing experiments were run with different numbers of threads, with OpenBLAS as the BLAS library. For comparison, **solve()** was run with R's built-in BLAS, and using the **gpuSolve()** function in **gputools**. A quad-core machine with a hyperthreading degree of 2 was used.

Here are the results, for a random 4000×4000 matrix:

platform	time
built-in BLAS	107.408
GPU	78.061
OpenBLAS, 1 thread	6.612
OpenBLAS, 2 threads	3.593
OpenBLAS, 4 threads	2.087
OpenBLAS, 6 threads	2.493
OpenBLAS, 8 threads	2.993

This is rather startling! Not only is OpenBLAS the clear champion here, but in particular it bests the GPU. The latter is substantially better than plain R, yes, but nowhere near what OpenBLAS gives us—even with just one thread. GPUs are invaluable for embarrassingly parallel problems, but obtaining excellent gains in other problems is difficult.

Also interesting is the pattern with respect to the number of threads. After moving past four threads, performance deteriorated. This pattern was confirmed in repeated trials, not shown here. This is not too surprising, as there are only four actual cores on the machine, even though each one has some limited ability to run two threads at once. To pursue this idea, these experiments were run on the 16-core machine, again with a hyperthreading degree of 2. Again performance appeared to saturate at about 16 cores.

12.8 Sparse Matrices

As mentioned earlier, in many parallel processing applications of linear algebra, the matrices can be huge, even having millions of rows or columns. However, in many such cases, most of the matrix consists of 0s.

This is quite common in data science applications. A good example is *market basket data.* Say we have n transactions, in each of which the consumer purchases one of more items among the s that the vendor has for sale. Each of the n records in the input file consists of a list of the items purchased, with each item coded by an ID in the range of 1 to s.

For various statistical/machine learning methods, we may wish to convert this data to an $n \times s$ matrix, filled with 0s and 1s. A 1 in row i, column j would indicate that item j was among those purchased in transaction i.

If there are an average of v items per transaction, then a proportion v/s of the matrix elements will be 1s. This figure is typically small, so our matrix will indeed be *sparse.*

In an effort to save memory, one can store sparse matrices in compressed form, storing only the nonzero elements. Sparse matrices roughly fall into two categories. In the first category, the matrices all have 0s at the same known positions. For instance, in *tridiagonal* matrices, the only nonzero elements are either on the diagonal or on subdiagonals just below or above the diagonal, and all other elements are guaranteed to be 0, such as

$$\begin{pmatrix} 2 & 0 & 0 & 0 & 0 \\ 1 & 1 & 8 & 0 & 0 \\ 0 & 1 & 5 & 8 & 0 \\ 0 & 0 & 0 & 8 & 8 \\ 0 & 0 & 0 & 3 & 5 \end{pmatrix} \tag{12.29}$$

Code to deal with such matrices can then access the nonzero elements based on this knowledge.

In the second category, each matrix that our code handles will typically have its nonzero entries in different, "random," positions, as in the market basket example. A number of methods have been developed for storing amorphous sparse matrices, such as the Compressed Sparse Row format, which we'll code in this C **struct**, representing an $m \times n$ matrix A, with k nonzero entries, for concrete illustration:

```
struct {
   int m,n;  // numbers of rows and columns of A
   // the nonzero values of A, in row-major order
   float *avals;
   int *cols;  // avals[i] is in column cols[i] in A; length k
   int *rowplaces;  // rowplaces[i] is the index in avals for
                    // the 1st nonzero element of row i in A
                    // (but the last element is k)
}
```

Since we're expressing matters in C, our row and column indices will begin at 0.

For the matrix in (12.29) (if we were not to exploit its tridiagonal nature, and just treat it as amorphous):

- **m,n:** 5,5
- **avals:** 2,1,1,8,1,5,8,8,8,3,5
- **cols:** 0,0,1,2,1,2,3,3,4,3,4
- **rowplaces:** 0,1,4,7,9,11

For instance, look at the 4 in **rowplaces**. It's at position 2 in that array, so it says that element 4 in **avals**—the third 1—is the first nonzero element in row 2 of A. Look at the matrix, and you'll see this is true.

Parallelizing operations for a sparse matrix A can be done in the usual manner, e.g. breaking the rows of A into chunks. Note, though, that there could be a load-balance issue, again addressable in ways we've used before.

Note that large sparse matrices may not need parallel computation in the first place, as the computation time depends on the number of nonzero elements, which may be manageable serially. On the other hand, keep in mind that if we multiply sparse matrices, the result may NOT be sparse, especially after doing several cumulative products, say with matrix powers.

As usual, one should search for good libraries first. There is PSBLAS, for instance, a version of BLAS for sparse matrices, running on top of MPI. For GPUs, there is the CUSP library.

12.9 Further Reading

There is an entire book titled *Hands-on Matrix Algebra Using R* (Hrishikesh Vinod, World Scientific, 2011), very useful in the context of this chapter.

The research literature on parallel numeric linear algebra is quite extensve. The reader would do well by beginning with *Numerical Linear Algebra on High-Performance Computers*, by J. Dongarra *et al*, SIAM, 1998.

The field of graph modeling has become quite popular in recent years. Many of the operations deal with matrices, typically large enough to require parallel computation. A good compendium of graph algorithms is *Practical Graph Mining with R*, by N. Samatova *et al*, CRC, 2014.

Chapter 13

Inherently Statistical Approaches: Subset Methods

A recurrent theme in this book has been that it is easy to speed up embarrassingly parallel (EP) applications, but the rest can be quite a challenge. Fortunately, there exist methods for converting many non-EP statistical problems to EP ones that are equivalent or reasonable substitutes.

I call this *software alchemy.* Such methods will be presented in this chapter, with the main focus being on one method in particular, Chunk Averaging (CA). Brief overviews will also be given of two other methods.

Let's set some notation. It will be helpful to have at hand the concept of a "typical rectangular data matrix," with n rows, i.e., n observations, and p columns, that is, p variables. Also, suppose we are estimating a population parameter θ, possibly vector-valued. Our estimator for the full data set is denoted by $\widehat{\theta}$, which we will often call the "full estimator."

13.1 Chunk Averaging

CA has been treated in specialized form by various authors since 1999. The general form presented here is adapted from my own research, *Software Alchemy: Turning Complex Statistical Computations into Embarrassingly-*

Parallel Ones, Norman Matloff, http://arxiv.org/abs/1409.5827

CA is quite simple: Say $\widehat{\theta}$ is generated by applying a function **g()** to our data. For instance, **g()** could be R's **glm()** function, computing a vector of estimated coefficients in logistic regression. Then, the steps of CA are the following:

(a) Break the data into r chunks of rows. The first $k = n/r$ observations comprise the first chunk, the next k observations form the second chunk, and so on.

(b) Apply **g()** to each chunk.

(c) Average the r results obtained in Step (b), thus producing our CA estimator $\widetilde{\theta}$ of θ.

Note that typically the result of **g()** is a vector, so that we are averaging vectors in Step (c).

If n is not evenly divisible by r, we can take a weighted average, with weights proportional to the chunk sizes. Specifically, let n_i denote the size of chunk i, and let $\widehat{\theta}_i$ be the estimate of θ on that chunk. Then the CA estimator is

$$\widetilde{\theta} = \sum_{i=1}^{r} \frac{n_i}{n} \, \widehat{\theta}_i \tag{13.1}$$

It is assumed that the observations are i.i.d. Then the chunks are i.i.d. as well, so one can also obtain the standard error, or more generally, an estimated covariance matrix. Let V_i be that matrix for chunk i (obtained from the output of **g()**). Then the estimated covariance matrix for $\widetilde{\theta}$ is

$$\sum_{i=1}^{r} (\frac{n_i}{n})^2 \, V_i \tag{13.2}$$

Like many procedures in statistics, this one has a tuning parameter: r, the number of groups.

13.1.1 Asymptotic Equivalence

Chunking is of course a time-honored method for parallelization, as we've seen throughout this book. But what makes CA different is that it is statistical in nature, and its usefulness derives from the fact that the CA estimator $\widetilde{\theta}$ is statistically equivalent to the full estimator $\hat{\theta}$.

It is easily proven that if the data are i.i.d. and the full estimator is asymptotically multivariate normal, then the CA method again produces an asymptotically multivariate normal estimator. And **most importantly**, the chunked estimator has the same asymptotic covariance matrix as the original nonchunked one.

This last statement, coupled with the fact that Step (b) is an embarrassingly parallel operation, implies that CA does indeed perform software alchemy—it turns a non-EP problem into a statistically equivalent EP one.

13.1.2 O(·) Analysis

Suppose the work associated with computing $\widehat{\theta}$ is $O(n^c)$, such as the $O(n^2)$ figure for our mutual outlinks example (Section 2.9). If the r chunks are handled in parallel, CA reduces the time complexity of an $O(n^c)$ problem to roughly $O(n^c/r^c)$ for a statistically equivalent problem, whereas a speedup that is linear in r would only reduce the time to $O(n^c/r)$.

If $c > 1$, then the speedup obtained from CA is greater than linear in r, which is called *superlinear*. This is a term from the general parallel processing field. When this occurs generally, the size of the effect is usually small, and is due to cache effects and the like. But in our statistical context, superlinearity will be commonplace, often with very large effects.

By the way, a similar analysis shows that CA can yield speedup even in the serial case, in which the chunks are handled one at a time. The time here will be $r\ O(\frac{n^c}{r^c}) = O(\frac{n^c}{r^{c-1}})$. So, for $c > 1$, CA may be faster than the full estimator even in uniprocessor settings.

Note too that CA can be helpful even in EP settings. Say our function **g()** is part of an existing software package. Even if the underlying algorithm is EP, recoding it for parallelization may be an elaborate undertaking. CA then gives us a quick, simple way to exploit the EP nature of the estimator. This will be illustrated in Section 13.1.4.3.

For some applications, though, our speedup may be more modest. As discussed in Section 3.4.1, linear (or nonlinear but still parametric) regression models present challenges. We'll see an example later in this section.

13.1.3 Code

Coding CA is a straightforward implementation of (13.1) and (13.2). It is available as the **ca()** function in my CRAN package, **partools**.

13.1.4 Timing Experiments

Let's look at speedups in a few examples.

13.1.4.1 Example: Quantile Regression

Here I used the **quantreg** package from CRAN in simulation experiments. There were m predictors, i.i.d. U(0,1), with the response being generated by

$Y = X_1 + ... + X_m + 0.2\ U$

with U again having a U(0,1) distribution. Sample sizes n were 25000 and 50000, with $m = 75$. Up to 32 threads were tried, on the machine described in the preface of this book.

The results, shown in Figure 13.1, are very illuminating. Strong speedups were obtained, especially in the case $n = 50000$, where the gain was superlinear. There seems to be no performance gain, and possibly even some decline, once we exceed 16 threads. Recall that this machine has 16 cores, though with hyperthreading degree 2, thus some potential for performance gain after 16 threads, but this was not observed in this application.

13.1.4.2 Example: Logistic Model

The data set here is the famous forest groundcover data from the University of California, Irvine Machine Learning Repository. Here we have seven different kinds of ground cover, with covariates such as Hillside Shade at Noon. The goal is to predict the type of ground cover based on remote sensing of the covariates. There are 500000 observations in the subset of the data analyzed here..

The timing experiments involved predicting Cover Type 1 from the first 10 covariates. Here is the code and timing results for the full data set:[1]

[1]Since the raw data is ordered, a random permutation is applied.

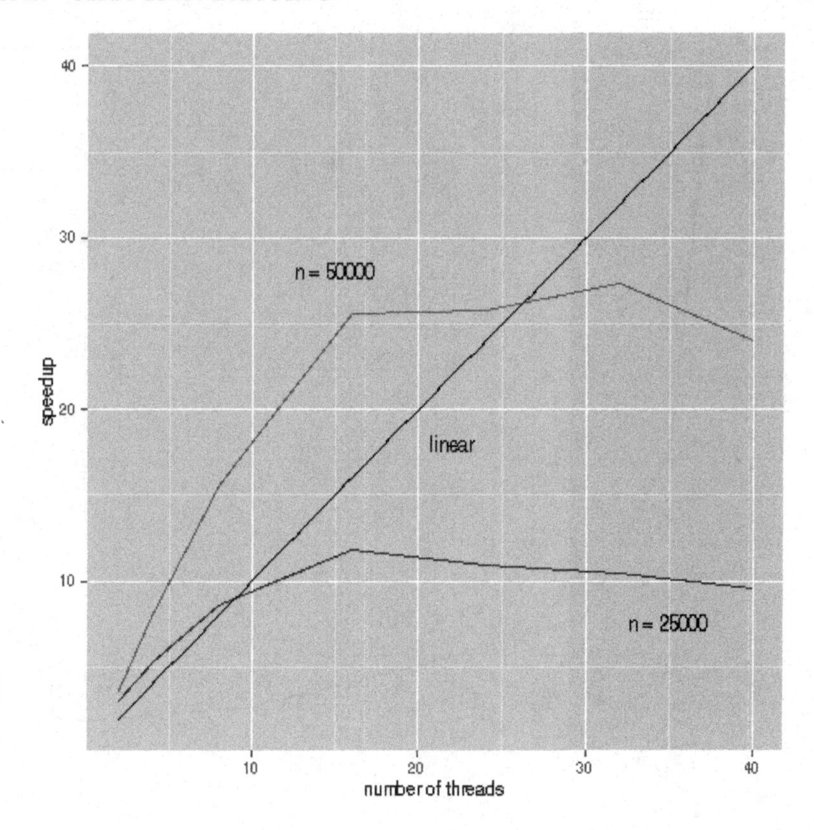

Figure 13.1: CA Performance, Quantile Regression

```
> forest <- read.csv("covtype.data")
# dummy for Type 1
> forest <- cbind(as.integer(forest[,55]==1),forest)
> forest <- as.matrix(forest)
> nrf <- nrow(forest)
> forest <- forest[sample(1:nrf,nrf,replace=FALSE),]
> system.time(g1 <- glm(forest[,1] ~ forest[,2:11],
    family=binomial))
   user  system elapsed
 40.174   1.754  41.977
```

Now compare that 41.977-second time to those obtained with CA run on the 16-core machine mentioned earlier, with 2, 4, 8, 12 and 16 threads,

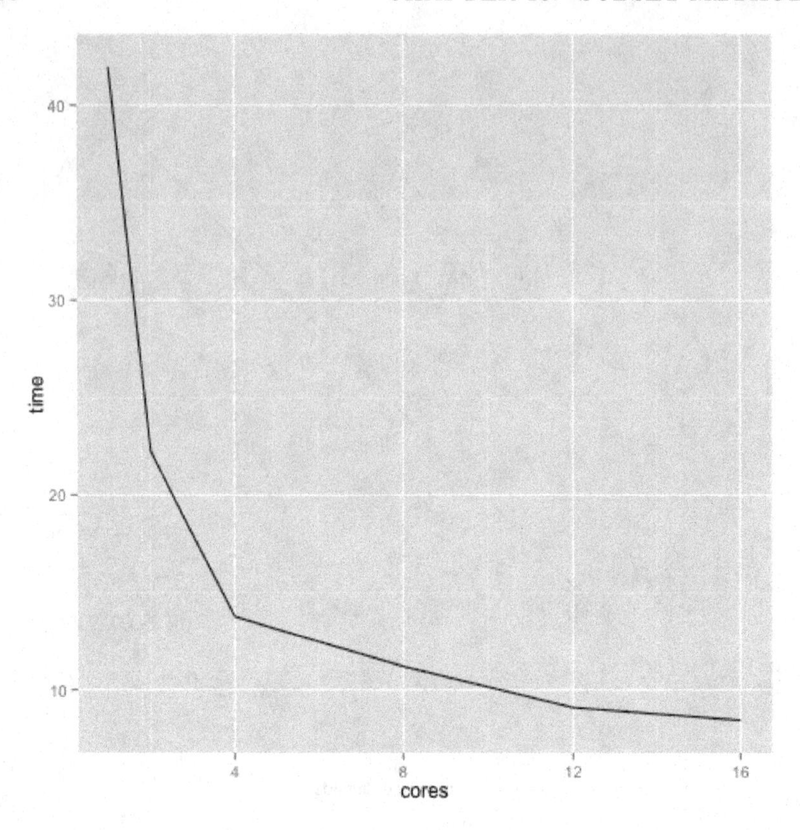

Figure 13.2: CA Performance, Forest Data

as seen in Figure 13.2. The speedup starts out linear, then becomes less dramatic but still quite good, especially in light of the factors mentioned earlier.

But what about the accuracy? The theory tells us that the CA estimator is statistically equivalent to the full estimator, but this is based on asymptotics. (Though it should be noted that even the full estimator is based on asymptotics, since **glm()** itself is so.) Let's see how well it worked here. Table 13.1 shows the values of the estimated coefficient for the first predictor variable, for the different numbers of cores (1 core again means the full-estimator case):

This is excellent agreement, even for the smallest chunk size.

cores	$\widehat{\beta}_1$
1	0.006424
2	0.006424
4	0.006424
8	0.006426
12	0.006427
16	0.006427

Table 13.1: Coefficient Estimates

13.1.4.3 Example: Estimating Hazard Functions

For a continuous distribution with density f and cdf F, the *hazard function* is defined as

$$h(t) = \frac{f(t)}{1 - F(t)}$$

The **muhaz** package on CRAN finds nonparametric estimates of h, for either censored or uncensored data. A variety of methods is offered, with the one used here employing kernel-based density estimation, with local bandwidth determination. The details are well beyond the scope of this book, but think of a histogram with varying bin widths, with the width of any particular bin being determined by the estimated characteristics of the density function in that region. Putting technical details aside, the key points are that (a) there is lots of computation to do and (b) the computation is embarrassingly parallel (in the uncensored case, which we have here).

At first, one might guess that since this procedure is embarrassingly parallel, we don't need CA. The latter method, after all, is aimed at converting non-EP problems to EP ones—so why use it on a problem that is already EP?

The answer is convenience. It would likely be quite inconvenient to rewrite the **muhaz** package for parallel computation. By contrast, if we apply CA to **muhaz**, we attain EP speed without any recoding.

So, let's see how well it does. The data set used was another famous one, the flight delay data, at `http://stat-computing.org/dataexpo/2009/`

`the-data.html`. The hazard function was estimated on the variable DepDelay (departure delay). NA values were removed from the data first. Here is the result for the full estimator:

```
> x <- read.csv("2008.500k.csv",header=TRUE)
> depdelay <- x$DepDelay
> depdelay <- depdelay[!is.na(depdelay)]
> library(muhaz)
> system.time(mhout <- muhaz(depdelay))
   user  system elapsed
 95.365   0.220  95.612
```

Then CA was tried, with 8 nodes:

```
> library(parallel)
> cls <- makeCluster(8)
> clusterCall(cls,function() library(muhaz))
...
> source("~/ChunkAveraging.R")
> system.time(
+    mhoutca <- ca(cls,depdelay,
+       ovf=function(zchunk) muhaz(zchunk)$haz.est[20],
+       estf=function(estoutchunk) estoutchunk)
+ )
   user  system elapsed
  0.244   0.020  14.418
```

Nice results, almost a seven-fold speedup.

13.1.5 Non-i.i.d. Settings

As noted, the usefulness of CA stems from the statistical equivalence of $\widetilde{\theta}$ and $\widehat{\theta}$. This in turn stems from the i.i.d. nature of our data. What about other settings?

The i.i.d. assumption is central to most of statistics. One seldom sees discussion of the implications of that assumption for day-to-day data analysis, but for CA there is an important reason to bring it up: For many data sets, the physical storage has been ordered in some way.

For example, suppose in a certain data set one of the variables is gender, say consisting of 5000 men and 5000 women. The data file may have been arranged so that the first 5000 records consist of the men and the second

5000 contain the data for the women. Suppose $r = 2$. Then the distribution in the first chunk is different from that of the second, i.e., the chunks are not identically distributed, and the CA theory doesn't hold.

Thus, if the analyst knows or suspects that the arrangement of the data is ordered in some way, he/she should first apply a random permutation to the n records. If the matrix **x** contains the original data, then one might run, say

```
x <- x[sample(1:n,n,replace=FALSE),]
```

13.2 Bag of Little Bootstraps

This intriguing method, the *Bag of Little Bootstraps* (BLB), is described in A Scalable Bootstrap for Massive Data, A. Kleiner, A. Talwalkar, P. Sarkar, M.I. Jordan, *Journal of the Royal Statistical Society*, Series B, 2013. To my knowledge, there is no publicly available code implementing BLB, and only an overview of the procedure will be presented here. (It is presumed that the reader has some familiarity with *bootstrap* methods.)

In BLB, one also looks at chunks of the data, as in the CA method, but the chunks are chosen randomly. We select s chunks of size b. For each chunk, we apply the standard bootstrap, taking r samples (with replacement) of size b. We then average over all chunks. As with CA, the authors prove that for i.i.d. data, BLB produces an estimator that is asymptotically equivalent to $\hat{\theta}$.

BLB has three tuning parameters: b, s and r. The above paper contains suggestions as to how to choose the values of these parameters.

13.3 Subsetting Variables

Consider a regression or classification setting. Instead of subsetting *observations*, i.e., rows of the data matrix, we might consider subsetting *predictor variables*, i.e., columns of the matrix. This a form of what is called *boosting*.

Say for example we have 50 predictor variables and we wish to do prediction of a binary outcome Y using a logistic regression model, via R's **glm()** function. Instead of one single call to **glm()** using all 50 predictors, and then applying the result to predict the Y values for whatever new data we encounter in the future, we might randomly select k pairs of predictors, and apply **glm()** to each pair of predictors. Then for each future data point, we would generate k predictions for that point, and simply use a majority rule to predict the new Y value. For instance, if the majority of our k predictions guess the new Y to be 1, then that would be our overall guess. Instead of using just pairs of predictors, we might use triples, or in general, m predictors for each small model fitted.

The motivation generally put forth for this approach is Richard Bellman's notion of the *curse of dimensionality*, which asserts that prediction becomes inordinately difficult in very high dimensions, that is with a very large number of predictors. We try to circumvent this by combining many predictions, each one in a low-dimensional space.

However, in our context here, one can view the above boosting method as a way to parallelize our operations. Consider linear regression analysis, for example, with p predictors. As discussed in Section 3.4.1, for fixed sample size, the work needed is $O(p^3)$, or $O(p^2)$, depending on the numerical method used. This time complexity grows more than linearly with p, so boosting may save us computation time. This is especially true since we fit our k models in parallel, an embarrassingly parallel setting.

Note that unlike the CA and BLB methods, boosting does not yield a statistically equivalent estimator. But it does save us time (and may reduce the chance of overfitting, accordingly to proponents).

This method has two tuning parameters: k and m.

13.4 Further Reading

For more details on the CA method, see "Software Alchemy: Turning Complex Statistical Computations into Embarrassingly-Parallel Ones," N. Matloff, to appear in the *Journal of Statistical Software*, 2015.

The Bag of Little Bootstraps is described in "A Scalable Bootstrap for Massive Data," A. Kleiner, A. Talwalkar, P. Sarkar and M.I. Jordan, *Journal of the Royal Statistical Society*, Series B, 2013.

Appendix A

Review of Matrix Algebra

This book assumes the reader has had a course in linear algebra (or has self-studied it, always the better approach). This appendix is intended as a review of basic matrix algebra, or a quick treatment for those lacking this background.

A.1 Terminology and Notation

A **matrix** is a rectangular array of numbers. A **vector** is a matrix with only one row (a **row vector** or only one column (a **column vector**).

The expression, "the (i,j) element of a matrix," will mean its element in row i, column j.

Please note the following conventions:

- Capital letters, e.g. A and X, will be used to denote matrices and vectors.
- Lower-case letters with subscripts, e.g. $a_{2,15}$ and x_8, will be used to denote their elements.
- Capital letters with subscripts, e.g. A_{13}, will be used to denote submatrices and subvectors.

If A is a **square** matrix, i.e., one with equal numbers n of rows and columns, then its **diagonal** elements are a_{ii}, i = 1,...,n.

A square matrix is called **upper-triangular** if $a_{ij} = 0$ whenever $i > j$, with a corresponding definition for **lower-triangular** matrices.

The **norm** (or **length**) of an n-element vector X is

$$\| X \| = \sqrt{\sum_{i=1}^{n} x_i^2} \tag{A.1}$$

A.1.1 Matrix Addition and Multiplication

- For two matrices have the same numbers of rows and same numbers of columns, addition is defined elementwise, e.g.

$$\begin{pmatrix} 1 & 5 \\ 0 & 3 \\ 4 & 8 \end{pmatrix} + \begin{pmatrix} 6 & 2 \\ 0 & 1 \\ 4 & 0 \end{pmatrix} = \begin{pmatrix} 7 & 7 \\ 0 & 4 \\ 8 & 8 \end{pmatrix} \tag{A.2}$$

- Multiplication of a matrix by a **scalar**, i.e., a number, is also defined elementwise, e.g.

$$0.4 \begin{pmatrix} 7 & 7 \\ 0 & 4 \\ 8 & 8 \end{pmatrix} = \begin{pmatrix} 2.8 & 2.8 \\ 0 & 1.6 \\ 3.2 & 3.2 \end{pmatrix} \tag{A.3}$$

- The **inner product** or **dot product** of equal-length vectors X and Y is defined to be

$$\sum_{k=1}^{n} x_k y_k \tag{A.4}$$

- The product of matrices A and B is defined if the number of rows of B equals the number of columns of A (A and B are said to be **conformable**). In that case, the (i,j) element of the product C is defined to be

$$c_{ij} = \sum_{k=1}^{n} a_{ik} b_{kj} \tag{A.5}$$

For instance,

$$\begin{pmatrix} 7 & 6 \\ 0 & 4 \\ 8 & 8 \end{pmatrix} \begin{pmatrix} 1 & 6 \\ 2 & 4 \end{pmatrix} = \begin{pmatrix} 19 & 66 \\ 8 & 16 \\ 24 & 80 \end{pmatrix} \tag{A.6}$$

It is helpful to visualize c_{ij} as the inner product of row i of A and column j of B, e.g. as shown in bold face here:

$$\begin{pmatrix} \mathbf{7} & \mathbf{6} \\ 0 & 4 \\ 8 & 8 \end{pmatrix} \begin{pmatrix} \mathbf{1} & 6 \\ \mathbf{2} & 4 \end{pmatrix} = \begin{pmatrix} 19 & 66 \\ 8 & 16 \\ 24 & 80 \end{pmatrix} \tag{A.7}$$

- Matrix multiplication is associative and distributive, but in general not commutative:

$$A(BC) = (AB)C \tag{A.8}$$

$$A(B + C) = AB + AC \tag{A.9}$$

$$AB \neq BA \tag{A.10}$$

A.2 Matrix Transpose

- The transpose of a matrix A, denoted A' or A^T, is obtained by exchanging the rows and columns of A, e.g.

$$\begin{pmatrix} 7 & 70 \\ 8 & 16 \\ 8 & 80 \end{pmatrix}' = \begin{pmatrix} 7 & 8 & 8 \\ 70 & 16 & 80 \end{pmatrix} \tag{A.11}$$

- If $A + B$ is defined, then

$$(A + B)' = A' + B' \tag{A.12}$$

- If A and B are conformable, then

$$(AB)' = B'A' \tag{A.13}$$

A.3 Linear Independence

Equal-length vectors X_1,...,X_k are said to be **linearly independent** if it is impossible for

$$a_1X_1 + ... + a_kX_k = 0 \tag{A.14}$$

unless all the a_i are 0.

A.4 Determinants

Let A be an $n \times n$ matrix. The definition of the determinant of A, det(A), involves an abstract formula featuring permutations. It will be omitted here, in favor of the following computational method.

Let $A_{-(i,j)}$ denote the submatrix of A obtained by deleting its ith row and j^{th} column. Then the determinant can be computed recursively across the kth row of A as

$$det(A) = \sum_{m=1}^{n} (-1)^{k+m} det(A_{-(k,m)}) \tag{A.15}$$

where

$$det \begin{pmatrix} s & t \\ u & v \end{pmatrix} = sv - tu \tag{A.16}$$

Generally, determinants are mainly of theoretical importance, but they often can clarify one's understanding of concepts.

A.5 Matrix Inverse

- The **identity** matrix I of size n has 1s in all of its diagonal elements but 0s in all off-diagonal elements. It has the property that $AI = A$ and $IA = A$ whenever those products are defined.
- The A is a square matrix and $AB = I$, then B is said to be the **inverse** of A, denoted A^{-1}. Then $BA = I$ will hold as well.

- A^{-1} exists if and only if its rows (or columns) are linearly independent.
- A^{-1} exists if and only if $det(A) \neq 0$.
- If A and B are square, conformable and invertible, then AB is also invertible, and

$$(AB)^{-1} = B^{-1}A^{-1} \tag{A.17}$$

A matrix U is said to be **orthogonal** if its rows each have norm 1 and are orthogonal to each other, i.e., their inner product is 0. U thus has the property that $UU' = I$ i.e., $U^{-1} = U$.

The inverse of a triangular matrix is easily obtained by something called **back substitution.**

Typically one does not compute matrix inverses directly. A common alternative is the **QR decomposition**: For a matrix A, matrices Q and R are calculated so that $A = QR$, where Q is an orthogonal matrix and R is upper-triangular.

If A is square and invertible, A^{-1} is easily found:

$$A^{-1} = (QR)^{-1} = R^{-1}Q' \tag{A.18}$$

Again, though, in some cases A is part of a more complex system, and the inverse is not explicitly computed.

A.6 Eigenvalues and Eigenvectors

Let A be a square matrix.[1]

- A scalar λ and a nonzero vector X that satisfy

$$AX = \lambda X \tag{A.19}$$

are called an **eigenvalue** and **eigenvector** of A, respectively.

[1]For nonsquare matrices, the discussion here would generalize to the topic of **singular value decomposition**.

- If A is symmetric and real, then it is **diagonalizable**, i.e., there exists an orthogonal matrix U such that

$$U'AU = D \tag{A.20}$$

for a diagonal matrix D. The elements of D are the eigenvalues of A, and the columns of U are the eigenvectors of A.

A different sufficient condition for A.20 is that the eigenvalues of A are distinct. In this case, U will not necessarily be orthogonal.

By the way, this latter sufficient condition shows that "most" square matrices are diagonalizable, if we treat their entries as continuous random variables. Under such a circumstance, the probability of having repeated eigenvalues would be 0.

A.7 Matrix Algebra in R

The R programming language has extensive facilities for matrix algebra, introduced here. Note by the way that R uses column-major order.

A linear algebra vector can be formed as an R vector, or as a one-row or one-column matrix.

```
> # constructing matrices
> a <- rbind(1:3,10:12)
> a
     [,1] [,2] [,3]
[1,]    1    2    3
[2,]   10   11   12
> b <- matrix(1:9,ncol=3)
> b
     [,1] [,2] [,3]
[1,]    1    4    7
[2,]    2    5    8
[3,]    3    6    9
# multiplication, etc.
> c <- a %*% b; c + matrix(c(1,-1,0,0,3,8),nrow=2)
     [,1] [,2] [,3]
[1,]   15   32   53
[2,]   67  167  274
> c %*% c(1,5,6)  # note 2 different c's
     [,1]
```

```
[1,]   474
> # transpose, inverse
> t(a)  # transpose
     [,1] [,2]
[1,]    1   10
[2,]    2   11
[3,]    3   12
> u <- matrix(runif(9),nrow=3)
> u
           [,1]       [,2]      [,3]
[1,] 0.08446154 0.86335270 0.6962092
[2,] 0.31174324 0.35352138 0.7310355
[3,] 0.56182226 0.02375487 0.2950227
> uinv
           [,1]      [,2]      [,3]
[1,]  0.5818482 -1.594123  2.576995
[2,]  2.1333965 -2.451237  1.039415
[3,] -1.2798127  3.233115 -1.601586
> u %*% uinv  # note roundoff error
             [,1]          [,2]          [,3]
[1,] 1.000000e+00 -1.680513e-16 -2.283330e-16
[2,] 6.651580e-17  1.000000e+00  4.412703e-17
[3,] 2.287667e-17 -3.539920e-17  1.000000e+00
> # eigenvalues and eigenvectors
> eigen(u)
$values
[1]  1.2456220+0.0000000i -0.2563082+0.2329172i
-0.2563082-0.2329172i

$vectors
              [,1]                  [,2]
[,3]
[1,] -0.6901599+0i -0.6537478+0.0000000i
-0.6537478+0.0000000i
[2,] -0.5874584+0i -0.1989163-0.3827132i
-0.1989163+0.3827132i
[3,] -0.4225778+0i  0.5666579+0.2558820i
0.5666579-0.2558820i
> # diagonal matrices (off-diagonals 0)
> diag(3)
     [,1] [,2] [,3]
[1,]    1    0    0
[2,]    0    1    0
```

```
[3 ,]    0    0    1
> diag((c(5,12,13)))
     [,1] [,2] [,3]
[1 ,]    5    0    0
[2 ,]    0   12    0
[3 ,]    0    0   13
```

Appendix B

R Quick Start

Here we present a quick introduction to the R data/statistical programming language. Further learning resources are listed at `http://heather.cs.ucdavis.edu/~/matloff/r.html`.

R syntax is similar to that of C. It is object-oriented (in the sense of encapsulation, polymorphism and everything being an object) and is a functional language (i.e., almost no side effects, every action is a function call, etc.).

B.1 Correspondences

aspect	C/C++	R
assignment	=	`<-` (or =)
array terminology	array	vector, matrix, array
subscripts	start at 0	start at 1
array notation	m[2][3]	m[2,3]
2-D array storage	row-major order	column-major order
mixed container	struct	list
return mechanism	return	return() or last value comp.
logical values	true, false	TRUE, FALSE
combining modules	include, link	library()
run method	batch	interactive, batch

B.2 Starting R

To invoke R, just type "R" into a terminal window. On a Windows machine, you probably have an R icon to click.

If you prefer to run from an IDE, you may wish to consider ESS for Emacs, StatET for Eclipse or RStudio, all open source. ESS is the favorite among the "hard core coder" types, while the colorful, easy-to-use, RStudio is a big general crowd pleaser. If you are already an Eclipse user, StatET will be just what you need.

R is normally run in interactive mode, with > as the prompt. Among other things, that makes it easy to try little experiments to learn from; remember my slogan, "When in doubt, try it out!"

B.3 First Sample Programming Session

Below is a commented R session, to introduce the concepts. I had a text editor open in another window, constantly changing my code, then loading it via R's **source()** command. The original contents of the file **odd.R** were:

```
oddcount <- function(x) {
   k <- 0 # assign 0 to k
   for (n in x) {
      if (n %% 2 == 1) k <- k+1  # %% is mod operator
   }
   return(k)
}
```

By the way, we could have written that last statement as simply

```
k
```

because the last computed value of an R function is returned automatically.

The R session is shown below. You may wish to type it yourself as you go along, trying little experiments of your own along the way.[1]

```
> source("odd.R")  # load code from the given file
```

[1]The source code for this file is at `http://heather.cs.ucdavis.edu/~matloff/MiscPLN/R5MinIntro.tex`. You can download the file, and copy/paste the text from there.

```
> ls()  # what objects do we have?
[1] "oddcount"
# what kind of object is oddcount?
> class(oddcount)
[1] "function"
# while in interactive mode, not inside a function,
# can print any object by typing its name; otherwise
# use print(), e.g. print(x+y)
> oddcount  # a function is an object, so can print
function(x) {
   k <- 0  # assign 0 to k
   for (n in x) {
      if (n %% 2 == 1) k <- k+1  # %% is mod operator
   }
   return(k)
}

# test oddcount(), but some traits of vectors first
> y <- c(5,12,13,8,88)  # c() is concatenate
> y
[1]  5 12 13  8 88
> y[2]  # R subscripts begin at 1, not 0
[1] 12
> y[2:4]  # extract elements 2, 3 and 4 of y
[1] 12 13  8
> y[c(1,3:5)]  # elements 1, 3, 4 and 5
[1]  5 13  8 88
> oddcount(y)  # should report 2 odd numbers
[1] 2

# change code (in the other window) to vectorize
# the count operation for much faster execution
> source("odd.R")
> oddcount
function(x) {
   x1 <- (x %% 2 == 1)
   # x1 a vector of TRUEs and FALSEs
   x2 <- x[x1]
   # x2 has elements of x that were TRUE in x1
   return(length(x2))
}

# try it on subset of y, elements 2 through 3
```

```
> oddcount(y[2:3])
[1] 1
> # try it on subset of y, elements 2, 4 and 5
> oddcount(y[c(2,4,5)])
[1] 0

# further compactify the code
> source("odd.R")
> oddcount
function(x) {
   length(x[x %% 2 == 1])
   # last value computed is auto returned
}
> oddcount(y)  # test it
[1] 2

# and even more compactification, making use of the
# fact that TRUE and FALSE are treated as 1 and 0
> oddcount <- function(x) sum(x %% 2 == 1)
# make sure you understand the steps that that
# involves: x is a vector, and thus x %% 2 is a
# new vector, the result of applying the mod 2
# operation to every element of x; then
# x %% 2 == 1 applies the == 1 operation to each
# element of that result, yielding a new vector
# of TRUE and FALSE values; sum() then adds them
# (as 1s and 0s)

# we can also determine which elements are odd
> which(y %% 2 == 1)
[1] 1 3

# now have ftn return odd count AND the odd
# numbers themselves, using the R list type
> source("odd.R")
> oddcount
function(x) {
   x1 <- x[x %% 2 == 1]
   return(list(odds=x1, numodds=length(x1)))
}
# R's list type can contain any type; components
# delineated by $
> oddcount(y)
```

```
$odds
[1]  5 13

$numodds
[1] 2

# save the output in ocy, which will be a list
> ocy <- oddcount(y)
> ocy
$odds
[1]  5 13

$numodds
[1] 2

> ocy$odds
[1]  5 13
> ocy[[1]]
[1]  5 13
# can get list elements using [[ ]] instead of $
> ocy[[2]]
[1] 2
```

Note that the function of the R function **function()** is to produce functions! Thus assignment is used. For example, here is what **odd.R** looked like at the end of the above session:

```
oddcount <- function(x)  {
   x1 <- x[x %% 2 == 1]
   return(list(odds=x1, numodds=length(x1)))
}
```

We created some code, and then used **function()** to create a function object, which we assigned to **oddcount**.

Note that we eventually **vectorized** our function **oddcount()**. This means taking advantage of the vector-based, functional language nature of R, exploiting R's built-in functions instead of loops. This changes the venue from interpreted R to C level, with a potentially large increase in speed. For example:

```
# 1000000 random numbers from the interval (0,1)
> x <- runif(1000000)
> system.time(sum(x))
```

```
   user  system elapsed
  0.008   0.000   0.006
> system.time({s <- 0;
for (i in 1:1000000) s <- s + x[i]})
   user  system elapsed
  2.776   0.004   2.859
```

B.4 Second Sample Programming Session

A matrix is a special case of a vector, with added class attributes, the numbers of rows and columns.

```
# rowbind() function combines rows of matrices;
# there's a cbind() for columns too
> m1 <- rbind(1:2,c(5,8))
> m1
     [,1] [,2]
[1,]    1    2
[2,]    5    8
> rbind(m1,c(6,-1))
     [,1] [,2]
[1,]    1    2
[2,]    5    8
[3,]    6   -1

# form matrix from 1,2,3,4,5,6, in 2 rows;
# R uses column-major storage
> m2 <- matrix(1:6,nrow=2)
> m2
     [,1] [,2] [,3]
[1,]    1    3    5
[2,]    2    4    6
> ncol(m2)
[1] 3
> nrow(m2)
[1] 2
> m2[2,3]  # extract element in row 2, col 3
[1] 6
# get submatrix of m2, cols 2 and 3, any row
> m3 <- m2[,2:3]
> m3
```

```
     [,1] [,2]
[1,]    3    5
[2,]    4    6

> m1 * m3 # elementwise multiplication
     [,1] [,2]
[1,]    3   10
[2,]   20   48
> 2.5 * m3  # scalar multiplication (but see below)
     [,1] [,2]
[1,]  7.5 12.5
[2,] 10.0 15.0
> m1 %*% m3  # linear algebra matrix multiplication
     [,1] [,2]
[1,]   11   17
[2,]   47   73

# matrices are special cases of vectors,
# so can treat them as vectors
> sum(m1)
[1] 16
> ifelse(m2 %%3 == 1,0,m2) # (see below)
     [,1] [,2] [,3]
[1,]    0    3    5
[2,]    2    0    6
```

The "scalar multiplication" above is not quite what you may think, even though the result may be. Here's why:

In R, scalars don't really exist; they are just one-element vectors. However, R usually uses **recycling**, i.e., replication, to make vector sizes match. In the example above in which we evaluated the express `2.5 * m3`, the number 2.5 was recycled to the matrix

$$\begin{pmatrix} 2.5 & 2.5 \\ 2.5 & 2.5 \end{pmatrix} \tag{B.1}$$

in order to conform with **m3** for (elementwise) multiplication.

The **ifelse()** function is another example of vectorization. Its call has the form

```
ifelse(boolean vectorexpression1, vectorexpression2,
   vectorexpression3)
```

All three vector expressions must be the same length, though R will lengthen some via recycling. The action will be to return a vector of the same length (and if matrices are involved, then the result also has the same shape). Each element of the result will be set to its corresponding element in **vectorexpression2** or **vectorexpression3**, depending on whether the corresponding element in **vectorexpression1** is TRUE or FALSE.

In our example above,

```
> ifelse(m2 %%3 == 1,0,m2) # (see below)
```

the expression `m2 %%3 == 1` evaluated to the boolean matrix

$$\begin{pmatrix} T & F & F \\ F & T & F \end{pmatrix} \tag{B.2}$$

(TRUE and FALSE may be abbreviated to T and F.)

The 0 was recycled to the matrix

$$\begin{pmatrix} 0 & 0 & 0 \\ 0 & 0 & 0 \end{pmatrix} \tag{B.3}$$

while **vectorexpression3**, **m2**, evaluated to itself.

B.5 Third Sample Programming Session

This time, we focus on vectors and matrices.

```
> m <- rbind(1:3,c(5,12,13))
> m
     [,1] [,2] [,3]
[1,]    1    2    3
[2,]    5   12   13
> t(m)  # transpose
     [,1] [,2]
[1,]    1    5
[2,]    2   12
[3,]    3   13
> ma <- m[,1:2]
> ma
     [,1] [,2]
```

```
[1,]    1    2
[2,]    5   12
> rep(1,2)  # "repeat," make multiple copies
[1] 1 1
> ma %*% rep(1,2)  # matrix multiply
     [,1]
[1,]    3
[2,]   17
> solve(ma,c(3,17))  # solve linear system
[1] 1 1
> solve(ma)  # matrix inverse
     [,1] [,2]
[1,]  6.0 -1.0
[2,] -2.5  0.5
```

B.6 The R List Type

The R **list** type is, after vectors, the most important R construct. A list is like a vector, except that the components are generally of mixed types.

B.6.1 The Basics

Here is example usage:

```
> g <- list(x = 4:6, s = "abc")
> g
$x
[1] 4 5 6

$s
[1] "abc"

> g$x  # can reference by component name
[1] 4 5 6
> g$s
[1] "abc"
# can reference by index, but note double brackets
> g[[1]]
[1] 4 5 6
> g[[2]]
```

```
[1] "abc"
> for (i in 1:length(g)) print(g[[i]])
[1] 4 5 6
[1] "abc"
```

B.6.2 The Reduce() Function

One often needs to combine elements of a list in some way. One approach to this is to use **Reduce()**:

```
> x <- list(4:6,c(1,6,8))
> x
[[1]]
[1] 4 5 6

[[2]]
[1] 1 6 8

> sum(x)
Error in sum(x) : invalid 'type' (list) of argument
> Reduce(sum,x)
[1] 30
```

Here **Reduce()** cumulatively applied R's **sum()** to **x**. Of course, you can use it with functions you write yourself too.

Continuing the above example:

```
> Reduce(c,x)
[1] 4 5 6 1 6 8
```

B.6.3 S3 Classes

R is an object-oriented (and functional) language. It features two types of classes, S3 and S4. I'll introduce S3 here.

An S3 object is simply a list, with a class name added as an *attribute*:

```
> j <- list(name="Joe", salary=55000, union=TRUE)
> class(j) <- "employee"
> m <- list(name="Joe", salary=55000, union=FALSE)
> class(m) <- "employee"
```

So now we have two objects of a class we've chosen to name **"employee"**. Note the quotation marks.

We can write class *generic functions*:

```
> print.employee <- function(wrkr) {
+   cat(wrkr$name,"\n")
+     cat("salary",wrkr$salary,"\n")
+     cat("union member",wrkr$union,"\n")
+ }
> print(j)
Joe
salary 55000
union member TRUE
> j
Joe
salary 55000
union member TRUE
```

What just happened? Well, **print()** in R is a *generic* function, meaning that it is just a placeholder for a function specific to a given class. When we printed **j** above, the R interpreter searched for a function **print.employee()**, which we had indeed created, and that is what was executed. Lacking this, R would have used the print function for R lists, as before:

```
# remove the function, to see what happens with print
> rm(print.employee)
> j
$name
[1] "Joe"

$salary
[1] 55000

$union
[1] TRUE

attr(,"class")
[1] "employee"
```

B.6.4 Handy Utilities

R functions written by others, e.g. in base R or in the CRAN repository for user-contributed code, often return values which are class objects. It is common, for instance, to have lists within lists. In many cases these objects are quite intricate, and not thoroughly documented. In order to explore the contents of an object—even one you write yourself—here are some handy utilities:

- **names()**: Returns the names of a list.
- **str()**: Shows the first few elements of each component.
- **summary()**: General function. The author of a class **x** can write a version specific to **x**, i.e., **summary.x()**, to print out the important parts; otherwise the default will print some bare-bones information.

For example:

```
> z <- list(a = runif(50),
   b = list(u=sample(1:100,25), v="blue sky"))
> z
$a
 [1] 0.301676229 0.679918518 0.208713522 0.510032893
 0.405027042 0.412388038
 [7] 0.900498062 0.119936222 0.154996457 0.251126218
 0.928304164 0.979945937
[13] 0.902377363 0.941813898 0.027964137 0.992137908
0.207571134 0.049504986
[19] 0.092011899 0.564024424 0.247162004 0.730086786
0.530251779 0.562163986
[25] 0.360718988 0.392522242 0.830468427 0.883086752
0.009853107 0.148819125
[31] 0.381143870 0.027740959 0.173798926 0.338813042
0.371025885 0.417984331
[37] 0.777219084 0.588650413 0.916212011 0.181104510
0.377617399 0.856198893
[43] 0.629269146 0.921698394 0.878412398 0.771662408
0.595483477 0.940457376
[49] 0.228829858 0.700500359

$b
$b$u
```

```
 [1] 33 67 32 76 29  3 42 54 97 41 57 87 36 92 81
 31 78 12 85 73 26 44
86 40 43

$b$v
[1] "blue sky"
> names(z)
[1] "a" "b"
> str(z)
List of 2
 $ a: num [1:50] 0.302 0.68 0.209 0.51 0.405 ...
 $ b:List of 2
  ..$ u: int [1:25] 33 67 32 76 29 3 42 54 97 41 ...
  ..$ v: chr "blue sky"
> names(z$b)
[1] "u" "v"
> summary(z)
  Length Class  Mode
a 50     -none- numeric
b  2     -none- list
```

B.7 Debugging in R

The internal debugging tool in R, **debug()**, is usable but rather primitive. Here are some alternatives:

- The RStudio IDE has a built-in debugging tool.
- For Emacs users, there is **ess-tracebug**.
- The StatET IDE for R on Eclipse has a nice debugging tool. It works on all major platforms, but can be tricky to install.
- My own debugging tool, **debugR**, is extensive and easy to install, but for the time being is limited to Linux, Mac and other Unix-family systems. See *http://heather.cs.ucdavis.edu/debugR.html*.

Appendix C

Introduction to C for R Programmers

The C language is quite complex, and C++ is even more so. The goal of this appendix is to give a start at a reading ability in the C language for those familiar with R.

C.0.1 Sample Program

```
// Learn.c

// inputs 5 numbers from the keyboard,
// squares and prints them

// include definitions needed for standard I/O
#include <stdio.h>

// function to square the elements of an array x,
// of length n, in-place; both arguments are of
// integer
int sqr(int *x, int n) {
   // allocate space for an integer i
   int i;
   // for loop, i=0,1,2,...,n-1
   for (i = 0; i < n; i++)
      x[i] = x[i] * x[i];
}
```

```
int main() {
   // allocate space for an array y of 10
   // integers, and a single integer i
   int y[10],i;
   for (i = 0; i < 5; i++)
      // input y[i]
      scanf("%d",&y[i]);
   sqr(y,10);  // call the function
   for (i = 0; i < 5; i++)
      printf("%d\n",y[i]);
}
```

Here is the compilation and sample run, using the **gcc** compiler:

```
$ gcc -g Learn.c
$ ./a.out
5 12 13 8 88
25
144
169
64
7744
```

C.0.2 Analysis

The comments explain most points, but a couple need detailed elaboration. First, consider line 20. Every C program (or other executable binary program, for that matter) is required to have a **main()** function, where execution will start.

As you see in line 23 and other places, every variable needs to be *declared*, meaning that we must request the compiler to make space for it. Since array indices start at 0 in C, this means we have to set up **y[0]** through **y[9]**. The compiler also needs to know the type of the variable, in this case integer.

The **scanf()** function has two arguments, the first here being the character string **"%d"**, which defines the *format*. Here we are specifying to read in one integer ('d' refers to "decimal number").

Things get more subtle in the second argument, where we see a major philosophical difference between C and R. The latter prides itself in (usually) not allowing *side effects*, meaning that in R one cannot change the value of an argument. The R call **sort(x)**, for instance, does NOT change **x**. What about C?

Technically, C doesn't allow direct changes to arguments either. But the key is that C allows—and makes heavy use of—*pointer variables.* For example, consider the code:

```
int u;
int *v = &u;
*v = 8;
```

The ampersand in **&u** means that the expression evaluates to the memory address of **u**. The asterisk in **v** tells the compiler that we intend **v** to contain a memory address. Finally, the line

```
*v = 8;
```

says, "Put 8 in whatever memory location **v** points to." This means that **u** will now contain 8!

Now we can see what is happening in that **scanf()** call. Let's rewrite it this way:

```
int *z = &y[i];
scanf("%d",z);
```

We are telling the compiler to produce machine code that will place the value read in from the keyboard to whatever memory location is pointed to by **z**—i.e., to place the value in **y[i]**. Convoluted, but this is how things work. We don't need it in line 29, as we are not changing **y[i]**.

The same principles are at work in lines 12 and 17. In the former, we state that **x** is a pointer variable. But how does that jibe with line 27? Why doesn't the latter write something like **&y**? The reason is that array variables are considered pointers. The simple expression **y** (without a subscript) actually means **&y[0]**.

If you are having trouble with this, console yourself with the fact that pointers are by far the most difficult concept that beguile novice C programmers (especially computer science students!). Just persist—you'll quickly get used to it.

C.1 C++

When the Object Oriented Programming wave came in, many in the C world wanted OOP for C. Hence, C++ (originally called "C with Classes").

The reader is referred to any of the excellent tutorials on C++ on the Web and in books. But here is a very brief overview, just to give the reader an inkling of what is involved.

C++ class structure is largely like R's S4. We create a new class instance using the keyword **new**, much like the call to **new()** we make in R to create an S4 class object.

As with S4, C++ classes generally contain *methods*, i.e., functions defined specific to that class. For instance, in our Rcpp code in this book, we sometimes have the expression **Rcpp::wrap**, meaning the **wrap()** function within the **Rcpp** class.

C++ continues C's emphasis on pointers. One important keyword, for example, is **this**. When invoked from some method of a class instance, it is a pointer to that instance.

Index

Note: Page numbers ending in "f" refer to figures. Page numbers ending in "t" refer to tables.

A

B

C

D

E

F

G

N

O

Q

R

S

T

Printed in the United States
by Baker & Taylor Publisher Services